W9-CIK-102

The Structure of Space

Physics and Humanities Series

The Structure of Space: *The Growth of Man's Ideas on the Nature of Matter*

by Joan Solomon

A HALSTED PRESS BOOK

The Structure
of
Space

*The Growth of Man's Ideas on the
Nature of Forces, Fields and Waves*

Joan Solomon

JOHN WILEY & SONS NEW YORK

ISBN 0470-81221-4

Library of Congress Catalog Card Number 73-8543

Published in the U.S.A.
by Halsted Press, a Division
of John Wiley & Sons, Inc.
New York

Printed in Great Britain

Contents

5

Introduction

Space is such a curious abstraction, in itself literally nothing at all, and yet just because of this it forms a very fertile ground for all kinds of speculation. In this decade it may call to mind the great achievements of the astronauts and the latest rocket probes towards the planets but in other ages it has suggested totally different ideas, sometimes almost purely religious, sometimes scientific, and sometimes philosophical. Always it presents the mind with an image of such vastness that the imagination is stretched to the limit to encompass it. For this reason the history of man's ideas about space is particularly fascinating because it demonstrates, better perhaps than any other subject, the changes in attitude and perception throughout the whole period of recorded human thought.

In this book will be found a mainly scientific selection of ideas about space although it is always impossible to isolate one line of inquiry from the total endeavour of the period. It took close on two thousand years from the time of Aristotle, for mankind to recognise the real emptiness of space. The medieval world had a phrase— 'horror vacuo'—which they attributed to a personified Nature but now it seems to us more applicable to their own timidity of imagination. To them the universe was neither infinite nor empty and the struggle to comprehend these two daring new concepts was painful and slow. Finally it was accomplished, a truly great achievement without which the whole of Newton's Theory of Gravitation could never have existed. This was the stage of understanding reached by the beginning of the eighteenth century and to a large extent it still remains the received opinion about space today. As far as Chemistry is concerned it does represent a kind of truth; space is almost completely devoid of matter—the perfect vacuum—but how dull this view appears when compared with the modern physicist's concept of a 'fabric' of forces linking the whole universe together!

It was the nineteenth century's researches into Electricity, Magnetism, and Light that began to populate the vast emptiness of space with a network of strains and oscillations which finally brought it to life. Michael Faraday is widely known as the inventor whose work led to the electric motor and the dynamo but he himself was always trying to describe to the world a personal feat of imagination that was far more revolutionary. He thought and wrote about the 'strains' and 'stresses' in space until they were even more real to him than were the charged objects and the magnets which he used to produce them. When Einstein began to formulate his new theory of gravity in General Relativity the idea of space as the seat of forces was to hand, ready to be formed into an even more powerful tool. He managed to increase the rôle of space still more until 'real' matter was reduced to a slight distortion of its curvature, a trivial local disturbance in its smooth and endless continuity. Einstein spent all the later years of his professional life trying to unite this view of space with the electro-magnetic strains of Faraday and Maxwell into one ambitious 'Unified Field Theory', but he never quite succeeded. It remains for some future generation of physicists to perceive the connection between gravitational curvature and electromagnetism so as to construct from them the next representation of force-filled space.

Another legacy that Relativity has given us is a new concept of time which is intimately linked with space. This idea has fired the imagination of scientist and layman alike until even children's television programmes are full of romances about time-travel and space flight. Mathematicians, cosmologists, and astronomers are all currently wrestling with the strange new properties of time though they seem to be throwing up more queries than they are answering. Still the challenge is basically as much to the power of human imagination as it is to the imposing difficulties of mathematics or the practical skills of astronomers.

These are some of the reasons why it is impossible to write about space without following the changing theories of Electricity, Magnetism, Gravity, and Light. Taken together they are the only measurable reality of space. This book has been provided with copious chapter and section headings so that the reader can pick out, if he wishes, those aspects of the subject that interest him most. Nevertheless he will always find, I hope, that the approach to scientific problems at any period in history is, to some extent, familiar because it is a part of the whole cultural climate of the times. With subjects so elusive and intriguing as space and time the whole of man's philosophy and aspirations must inevitably be drawn into his continual search for knowledge.

I

Soul, Movement, and the Aetherial Spheres

The most impressive mysteries of the physical world, which touch human imagination most deeply, are, it seems to me, the unseen forces which pierce the vast emptiness of space and influence distant worlds. Ever since childhood I have tried to picture our own familiar Earth, isolated in dark space, as it is pulled by a mysterious attraction into a lonely orbit round the distant fire-ball that we call the sun. No amount of scientific knowledge can, for a moment, match the immense wonder of the image! In this century it has inspired numbers of Science Fiction stories about voyages through space and time, powerful force-fields and deadly radiations. In earlier times such powers were attributes of gods alone and organised religions exploited this perennial awe of space. Now, in a less believing age, small children are still fascinated by simple magnets and will play with them for hours 'feeling' the invisible barrier of repulsion between them and the strange force that can lift without contact. This is an account of some of the 'powers' that reach through space creating a reality out of the void.

THE CREATION OF LIGHT

The blackest thing in the world is emptiness. The darkness of a deep chasm or well is complete and unnerving but the darkness of the night must have been a thousand times more frightening to primitive man. In the earliest myths it is always the Sun and its light which is the foremost power or god on Earth. Nowhere is this conquering of the fearful dark void better shown than in the opening verses of the Bible:

> In the beginning God created the heaven and the earth. And the earth was waste and void; the darkness was upon the face of the deep; and the spirit of God moved upon the face of the waters.
> And God said 'Let there be light', and there was light.
> And God saw the light that it was good.

This welcome piercing of the primeval darkness by the creation of light is to be found in most mythologies from China and India to Greece. The hymns which Hesiod wrote more than 2,700 years ago contain a myth of creation in which Chaos (literally 'yawning' or 'gaping') represented the frightening emptiness that exists where no light is.

> From Chaos were born Erebos (dark regions) and black Night; and from Night in turn Bright Sky (Aether) and Day, whom Night conceived and bore in loving union with Erebos. (*Theogony*— Hesiod.)

So the first power to pierce through the darkness of empty space was light itself in ancient mythology as it is again each day at dawn.

Just before the beginning of written philosophy a religious cult existed in the eastern Mediterranean embodying the oriental idea of life after death, the symbol of the eternal egg, and the worship of Apollo. This was the cult of Orpheus and its followers who sang of the creation of gods at the beginning of the world. These gods were not only anthropomorphic beings but also mystic ideas of space and time, of light and darkness. Chronos, or Time, is portrayed in Greek art of the seventh century BC as a multi-headed winged snake, and from him sprang Chaos (the Void), Aether (the bright upper air), and Erebos (the dark regions).

From the shining egg when it burst, came forth light, Heaven and Earth, and hence all living things. This, it is thought, was part of the contents of the early Rhapsodies which were recited in Orphic rites for centuries before they were written down—images that nourished the fervour of the participants with these grand concepts of Space and Time which have never lost their power.

However, mythical ideas of Void, Time, and Light were not enough to stimulate some early Einstein towards a modern view of space because they were too static. It needed forces stretched out across the emptiness causing power and movement to bring it to life and this was the very problem that was to perplex scientific thought from that time to the present day.

THE MOVING 'SOUL' OF THINGS

There was plenty of movement to be seen in Nature but at first it was all attributed to divine power. In heroic times even the flight of a spear seemed to be as much in the hands of fate as in the skill of the thrower. Homer's tales of the battles around Troy are full of gods and goddesses guiding (or mis-guiding!) arrows in mid-air. When men first tried to free their explanations from this random divine intervention and to build their own system of natural philosophy they found a bewildering number of different kinds of movement to explain. Falling trees, the circling Moon, rising smoke, and the driving rain seem to have so little in common that no universal cause was imaginable at that stage. Neither could the intervening space be responsible for these various movements. It seemed to be the objects themselves that burst into motion and the first philosophers could only exclaim with Thales—'All things are full of gods!'

Such a phrase may have seemed a little naïve and pantheistic even in those early days (c 500 BC). Perhaps it was too reminiscent of nymphs in every tree and sprites by every spring! At all events the words were changed and the 'gods' were dropped in favour of the 'soul' in things. From Thales to Plato and Aristotle, two

13

hundred years later, the active principle inside material objects which caused their movement was their *anima* or soul. This made all objects 'alive' to some extent with their own motive force and, like the forsaken gods themselves, both immortal and divine through their power to move.

> For they all seem to assume that movement is the distinctive character of the soul, and that everything else owes its movement to the soul, which they suppose to be self-moved, because they see nothing producing movement which does not itself move. (Aristotle on earlier philosophers' views from *On the Soul*.)

The ancient world also knew a little about magnetism and electricity. Lodestone—a natural magnet—is found to this day in Asia Minor and the very word 'magnet' is said to derive from the town of Magnesia on the Aegean coast. Lodestone was known as 'magnesian stone' in England up till Elizabethan times and many far-fetched stories were told of its wonderful powers. Sailors believed that magnetic islands existed which could pull the nails out of any boat venturing too near so that it fell apart in the water; they sometimes built their ships with wooden pegs instead of nails to avoid this terrible hazard! A lodestone in the hand could be seen to attract pieces of iron towards itself yet this was never used by the Greeks as an analogy to the falling of a stone to the ground. It was a perfect example of power pulling at an object some distance away but it remained only a marvel in the ancient world and a tribute to the strength of the lodestone's soul!

Primitive electricity was seen in the same way. It was known only that rubbed amber could mysteriously pull pieces of fluff or small seeds towards itself just as rubbed gramophone records attract dust. The Greek word for amber is 'electron' and though it was to name a whole new science the property seems to have aroused little interest at the time. Many centuries were to pass before it was demonstrated that any other dry substance could produce this

effect or even that the phenomenon was clearly different from magnetism.

Aristotle did produce a theory of what we should call gravity, but it contained no reference to either force or space. His explanation was *not* in terms of what pulls at heavy objects but what 'ideal' arrangement the *soul* in them was attempting to achieve. This is a very typical approach for the times in which he lived, only a few years after Plato had written of his famous ideal 'Republic' and when the perfection of pure mathematics was more admired

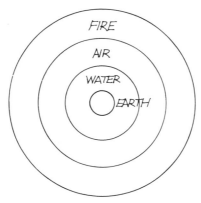

Ideal arrangement of the elements in the sub-lunar regions

than practical measurements. Aristotle believed that each of the four elements had its *proper place* in the Universe: earth at the centre, then water, air, and lastly fire. Earth and water were *heavy* because they tried to move downwards towards their places, but fire and air were *light* because they had a tendency to move upwards to their places.

THE EARTH IS ROUND

Since all earthly matter either has moved or is moving towards a centre this theory would lead to the Earth itself being a sphere at

the centre of the Universe. This great imaginative leap pleased Aristotle because he believed the sphere to be a 'perfect' shape but he also adduced evidence from observation to prove his point.

> Further proof is obtained from the evidence of the senses. (I) If the earth were not spherical, eclipses of the moon would not exhibit segments of the shape they do. . . . Thus if the eclipses are due to the interposition of the earth, the shape must be caused by its circumference, and the earth must be spherical. (II) Observation of the stars also shows not only that the earth is spherical but that it is of no great size, since a small change of position on our part southward or northward visibly alters the circle of the horizon, so that the stars above our heads change their position considerably. . . . For this reason those who imagine that the region around the Pillars of Heracles joins on to the regions of India, and that in this way the ocean is one, are not, it would seem, suggesting anything utterly incredible. They produce also in support of their contention the fact that elephants are a species found at the extremities of both lands, arguing that this phenomenon at the extremities is due to communication between the two. Mathematicians who try to calculate the circumference put it at 400,000 stades. (from *On the Heavens*—Aristotle.)

Aristotle's arguments from the eclipse of the moon when the earth's shadow can be seen to be circular, and his deductions from the constellations visible at different latitudes are clear and convincing proof of the Earth's shape. But his 'evidence' from the exist-

Earth's shadow thrown on the Moon during an eclipse

ence of elephants in north Africa and north India can only raise a smile today when we know how very much greater the sphere of the earth really is. (400,000 stades—the very first estimate of the earth's circumference—is equivalent to about 9,987 miles, only about two-fifths of its actual size.) This passage is said to be the one which inspired Christopher Columbus, some eighteen centuries later, to set out on his historic voyage westwards round the world to the Indies.

THE HEAVENLY SPHERES

The Sun, the Moon, the stars, and the planets, however, move neither up nor down but forever in great circles through the heavens. To account for this Aristotle postulated a fifth element— the quintessence—which was neither heavy nor light but whose soul moved it round and round the skies. For this substance he chose the old word 'aether' because this means 'runs always' and so it expressed for him and for all the Greeks an innate belief in the divine and eternal nature of the celestial bodies. In all cultures whether the stars were gods or not, circular shape and movement has always symbolised the ceaseless running of time, from prayer-wheels to horoscopes, from the oriental egg of eternity to the orb of the reigning monarch.

If the passage of time was continuous and eternal to Aristotle, space was definitely not. Since the outermost sphere of the total Universe was made of solid aether and rotated constantly it followed that the sphere must be of finite size otherwise infinitely distant parts of it would have to move with infinite speed to keep up, and this was clearly considered impossible. So a round 'tight, little Universe' was pictured in the ancient world with no unimaginable distances in it. Constructed of an incorruptible heavenly aether it encircled our earth in a nest of revolving concentric spheres. The last, all-enveloping, sphere carried the fixed constellations of stars, and ancient astronomers could calculate its size as 'the radius of the

Universe'—a curious phrase to our ears—with meaning, if little accuracy with regard to the distance of the stars.

Plato began the enumeration of the heavenly spheres with only eight which carried the moon, sun, five planets, and the fixed stars. As these were eternal bodies he thought they should revolve 'perfectly', at a steady speed, but it was well known that the facts were quite otherwise—some planets make clear 'loops' in the sky and none, not even the moon, move as simply as perfection would demand. Characteristically the Greeks did not try to amend or

Medieval illustration of the heavenly spheres

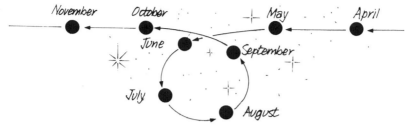

The loop of Mars in the night sky during 1939

change their theory; they held to it but set themselves a mathematical problem: how could all these observed variations be accounted for so that:

(a) there was steady rotation of all the heavenly bodies,

(b) the spheres all revolved around the earth, and

(c) the speed of rotation diminished regularly from the fast-moving outer sphere inwards toward the stationary Earth at the centre?

The Greeks were superb mathematicians and they almost solved the problem by adding more and more spheres until nearly all the

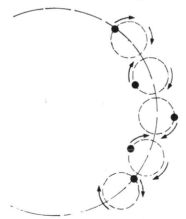

'Spheres rolling on spheres.' The epicycles of Ptolemy

known movements were accounted for. By the first century AD Claudius Ptolemy effectively completed the system by rolling spheres round spheres but the total number used in the calculation had now reached eighty! This was the most accurate and authoritative system of the ancient world and lasted, with some vicissitudes, until the sixteenth century. The spheres were held to be made of solid crystal and each one was geared to the one beyond it so that the whole system was spun round by the outermost sphere— the *Primum Mobile* or motive force of the Universe.

2

The Long Pause

It is true that there was a Dark Age in astronomy when the clock was turned back but this was due not so much to the barbaric horde that overran the Empire and sacked Rome itself as to a literal interpretation of the Bible that was insisted upon by the early Church fathers. They were anxious to make a fresh start in all branches of philosophy based only on the new faith and to throw out all previous secular knowledge since, in their view, 'the virtues of the heathens were but splendid vices'. According to the Bible the Earth was flat and had 'four corners', the heavens had been 'stretched out as a tent' and there were 'waters above the earth' as well as below it. This was divinely revealed knowledge and, as such, far surer than mere human reason! Aristotle's and Ptolemy's original works were lost to Europe at this time but there were enough references to them in later Roman writers to keep the old theories just alive during those long dark centuries. The orthodox beliefs of the times can be read in the four books of *Christian Topography* written about AD 540. The very title of the first book shows how firmly faith attempted to subjugate free thought: 'Against those who, while wishing to profess Christianity think and imagine like the pagans that the heaven is spherical.' The Earth, because it is heavy, is now obviously at the bottom of the Universe and is shaped exactly like the tabernacle which the Children of Israel carried through the wilderness. Individual angels carry the stars and planets along their appointed paths by God's unquestionable command. However, there were a few Christian scholars, like the Venerable Bede (673–735), who dared to quote the ancient view of the Universe. Eventually, by the tenth century, both the Earth and the heavens were again seen to be round although it took a longer time for the popular imagination to recapture the old ideas and make them real.

THE 'CRYSTAL SPHERES RETURN'

Although it had been the Christian Church of the fourth century which had first opposed the Aristotelean system of the Universe it was that same church which, in the thirteenth century, finally welcomed it back again. When the ancient books, retranslated into Latin, became available in Europe again it was the theologians who studied them so eagerly and taught their theories. They clothed the pagan 'spheres of aether' with Christian meaning and populated them with Christian angels. St Thomas Aquinas, who did more than any other single man to advocate the works of Aristotle, used the movement of the spheres as his first and foremost proof of God's very existence.

By the time that Dante lived and wrote *The Divine Comedy* the spheres and epicycles of the ancient mathematical system had become so familiar a part of popular belief that they could be the vehicle of poetry. In the third part of his great work—'Paradise'— Dante was led upwards by Beatrice through each of the nine celestial spheres in turn until they reached the *Primum Mobile* whose movement turned the whole universe by divine love. Beatrice, bathed in joy and radiance, explained to Dante how this god-like realm revolves the slower inner spheres and is the actual measure of time itself.

> The nature of the universe which stills
> The centre and revolves all else, from here,
> As from its starting-point, all movement wills.
>
> This heaven it is which has no other 'where'
> Than the Divine Mind; 'tis but in that Mind
> That love, its spur, and the power it rains inhere.

This circle's motion takes no measurements
From other spheres beneath, but theirs computes,
As two and five of ten are dividends.

As in a plant-pot, then, time has its roots
Herein, and where the other heavens trace
Their course, thou mayst behold its shoots.

This glorious vision of the heavens continued to inspire religious wonder for many generations. The hymn that Milton wrote three hundred years later, and which is still sung today, contains the same imagery:

Ring out ye Crystall sphears
Once bless our human ears. . . .

And with your ninefold harmony
Make up full consort to th' Angelike symphony.

Yet, poetry aside, such a picture of the Universe does lack something in true vastness and mystery. Despite their aetherial and eternal nature the spheres were, after all, no more than a great, interlocking machine. Ptolemy's Universe had to be spun by the hand of God, or by his angels, much as a child might spin a top or a set of cog-wheels. The rules of earthly force and motion did not apply 'up there' and there could be no questioning of speeds of the planets so securely embedded in their revolving spheres. To the medieval mind the idea of a solitary object being moved remotely through a void would have seemed, by contrast, the wildest, most unconvincing fantasy!

'NATURE ABHORS A VACUUM'

Empty space was inconceivable from Aristotle to Aquinas. Beyond the outermost sphere of their cosmic model there was nothing;

neither space nor time. In the total plan of God this region was not the 'proper place' of any substance—earthly or celestial—so it simply could not exist. Philosophers of those times teased their minds with the metaphysical problem: 'Is it possible to raise one's hand while standing on the edge of the Universe?' The only school of thought that could visualise a true void was that of the ancient Atomists. To them matter consisted of myriads of tiny atoms moving and colliding ceaselessly in a vacuum and indeed they held that this space was essential to provide room for the atoms to move into. At first this was not a popular theory. To talk about the 'existence of nothing' offended the logical Greek mind, so other ways of picturing motion had to be invented.

Plato tried to solve the problem of movement through a continuous medium by a kind of circular thrust in which the air parted in front of the object, skirted round to the back and then pushed it forward! This curious mechanism was used by Platonists to explain

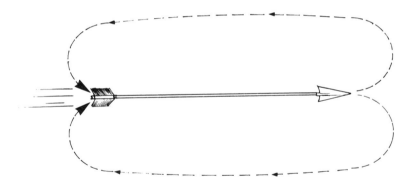

the flight of an arrow, the attraction of a nail to lodestone and the action of medical 'cupping glasses' used to draw out the infection from a wound or abscess. It is hard to see how such an explanation could convince men more practical than Plato unless they were very devoted followers. However, it did become recognised that the urge to fill up any space left by retreating air was a real and useful force.

The next centre of science was in Alexandria and it was here that the practical sciences began. The great Archimedes had laid the basis of a new science of mechanics in the second century BC, and he was followed by others including Hero of Alexandria who designed working models to illustrate the forces of air, steam, and water power. These men were inventors of little more than toys— small pumps, syphons, and engines. By the action of heat or suction they could remove air and make a partial vacuum but this, they found, was always filled in rapidly and with considerable power.

Nothing was known of the pressure of atmospheric air until the seventeenth century so the existence of this force of suction could be seen only as a proof of the 'unnaturalness' of a void. Throughout the Middle Ages it was held that 'Nature abhors a vacuum' in spite of the logical difficulty of maintaining that an empty space could, of itself, pull at anything at all! Galileo actually used the height of water that a pump could raise as a measure of the 'force of a vacuum'. Finally when the barometer and air pump were invented not only was the pressure of the atmosphere demonstrated but the emptiness of a vacuum could be observed and studied in a laboratory. Until that time, however, the very idea of a void was not only held to be philosophically unsound but felt to be abnormal and unnerving like the dark edge of a hole.

THE LIGHT OF GOD

Fear of the unknown darkness is such a primitive emotion that it is not surprising to find Light being widely used as a symbol of divine power by the early Christian Church just as Plato had earlier used it for intellectual power. Its radiance became a continual analogy for God's love and omnipotence in medieval theology stemming from the writings of St Augustine in the fourth century AD, where it was a literary image of great beauty and subsequent authority.

. . . if we understand the angels' creation aright herein, they are made partakers of that eternal Light, the unchangeable Wisdom of God, all-creating, namely the only begotten Son of God, with whose light they in their creation were illuminate, and made light, and called day in the participation of the unchangeable light and day, that Word of God by which they and all things else were created. (from *The City of God*—St Augustine, Book XI, Ch VIII.)

The ancient world had made few guesses as to the nature of light, being divided even as to whether it issued from the bright

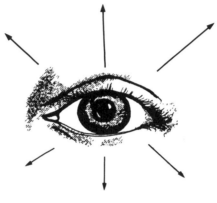

A 'beaming' eye

object or came in beams from the observer's eye and reached out 'feeling' the luminous object as a blind man feels with a walking stick! Light was a mystery and a challenge to science from the beginning because, unlike primitive magetism, electricity, or even gravity, its connection with the sun and the stars was unquestionable. The Atomists characteristically thought of both light and heat as tiny particles, Plato called it an 'emanation' but for Aristotle, to whom the universe was full of substance, light was propagated through the transparent medium as a 'potentiality' of form—like an impression of an object is transmitted through wax. The formation of shadows gives some idea of the practical basis of the action of

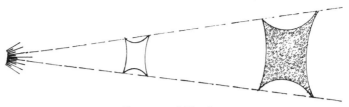

Shape and Shadow

distant, immaterial light. This theory, subtle and mysterious, appealed to the imagination of medieval philosophers. The laws of optics could be studied through geometry but the ideas of light and space were so interwoven that it became more a subject for meta-physics than for mathematics. By the thirteenth century there was speculation that light itself had been the cause of creation and movement as it radiated through the Universe impressing form on chaos. Historically this was the first theory of 'action at a distance' and it included heat rays, astrological influence, and mechanical effects as well as illumination. Most writers believed that the passage of light was instantaneous but some at least imagined that it moved in pulses or compressions like sound waves travel from the prongs of a vibrating tuning fork. This may seem a thoroughly modern and scientific approach but, in fact, the early theories of light were always loaded with wonder and religious awe; the stars, the Moon,

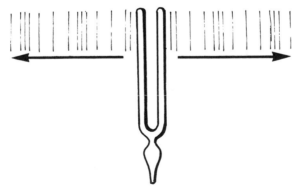

but most of all the Sun itself, radiated, in their light and heat, a God-like influence on all creation. In 1543 when Copernicus published his famous theory that the earth and the planets revolve around the sun the same mystical feeling was present. Although his argument was logical and his mathematical computations were impressively accurate the passage where Copernicus wrote of the sun was freely lyrical:

> In the middle of all sits Sun enthroned. In this most beautiful temple could we place this luminary in any better position from which he can illuminate the whole at once? He is rightly called the Lamp, the Mind, the Ruler of the Universe; Hermes Trismegistus names him the Visible God, Sophocles' Electra calls him the All-seeing. So the Sun sits as upon a royal throne ruling his children the planets which circle round him. (from *De Revolutionibus*—Copernicus.)

3
Revolution in Astronomy

Copernicus started a revolution in cosmology although, like most of such startling events, there had been premonitions of it in the centuries before. As early as the third century BC there had lived in Samos a very great astronomer, Aristarchus, who had dared to measure the sizes and distances of the Sun and Moon and had suffered a charge of impiety in consequence. His method was simple

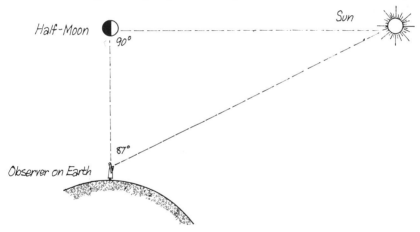

Aristarchus' method of comparing the distances of the Sun and the Moon

and although his results were grossly inaccurate by our standards he did appreciate how much larger the Sun was than the Moon or Earth. This estimate of celestial sizes was, in itself, very bold for the times but Aristarchus went further. Since the Sun was so massive he thought it more likely that the Earth revolved while the Sun re-

mained ponderously at rest. This was Aristarchus' blasphemy—
'moving the hearth of the Universe'—and it is hardly surprising
that he had few followers. However, when the ancient books became accessible once more in Europe (in the fourteenth and fifteenth
centuries for these more obscure writers) the seeds of the Copernican revolution lay ready to be sown.

Others in the century before Copernicus had toyed with the idea

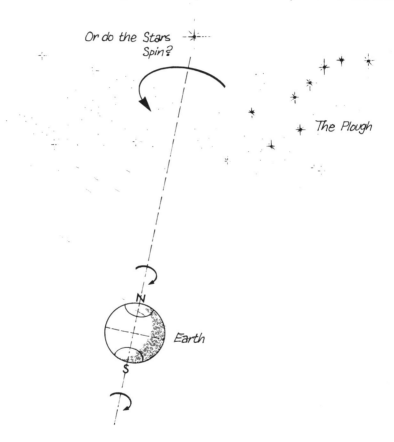

*Rotation of the Earth from West to East or of the stars from East
to West*

of a moving Earth, realising that, since all motion is relative, the appearance of motion in the Sun and stars would be identical whether they revolved round the Earth or it spun on its own axis. Nicolas of Cusa maintained that such axes of rotation were the only fixed realities in the Universe. 'There is no centre but the pole, which is God ever to be blessed.' But such an argument was too vague and metaphysical to stand against the calculated spheres and epicycles of Ptolemy. A mathematician of the first rank was required to launch the astronomical revolution and Canon Nicholas Copernicus was such a man.

FEWER SPHERES AND SIMPLER MATHEMATICS

It was just those very mathematical devices, complicated and meaningless, that first led Copernicus to distrust the Ptolemaic system. By his own account he then turned to the ancient philosophers to seek an alternative and found his own half-formed ideas already discussed in their works. This reassured him for he was not by nature an initiator and he found, to his great satisfaction, that a sun-centred Universe simplified the mathematics of the rotating spheres most elegantly. The simple yearly revolution of the Earth around the Sun gave rise to all the 'loops' in the paths of the planets without recourse to epicycles and equants (eccentric circles). It also explained the variation in apparent brightness as viewed from the Earth. The daily rotation of the Earth on its axis reduced the Sun to rest, simplified the orbits of the planets and, best of all, kept the fixed stars stationary instead of speeding improbably round their immense circle once in every twenty-four hours. However, he could not make do without a number of epicycles since he refused, like Aristotle, to envisage any celestial motion that was not strictly circular. He also kept all the crystalline aetherial spheres to carry the planets on their new courses.

In spite of these familiar landmarks Copernicus anticipated serious

Long exposure photograph of star-trails in the region round the Pole show unmistakably their rotation—or is it that of the Earth?

criticism on at least two points. First, no one could easily adjust to the idea that our Earth, the undisputed centre of local gravitational fall, was really spinning round at such a speed each day from west to east. They felt sure that such rotation would throw off all objects on the surface of the world, that a gale would blow continuously from the east and that a stone dropped from the top of a tower would fall well behind its base. It all seemed totally contrary to everyday experience, but the counter-arguments were uncertain since the whole basis of Mechanics was still to be formulated. At best this rotation could not be either proved or disproved. The second motion of the Earth, in a great circle round the Sun, should, however, have been detectable in the apparent movements of the stars. This effect is called 'parallax' (like the landscape apparently drifting backwards as seen from a moving train-window). Copernicus, like Aristarchus before him, was bound to admit that it was, as yet,

Argument about dropping a stone from a high tower

The tower moves on due to the rotation of the Earth—so that the stone is left behind

unobservable. To save his theory he had to conclude that this parallax existed but was too small to measure owing to the immense distance away of the fixed stars. (It needed, in fact, nineteenth-century telescopes to detect any parallax in even the nearest stars.) The Universe had had to expand alarmingly so that the whole yearly path of the Earth was little more than a point in comparison to the radius of the outer sphere of stars. Taken altogether Copernicus' system seemed to run grossly against the 'common-sense' view of things and could only seem laughable to most men at this time. But Copernicus was not deeply concerned about the new feats of imaginative understanding required of the ordinary man—he proposed the theory for its more economical solution of the celestial

33

movements and, as such, he was addressing his ideas principally to those mathematicians capable of appreciating this:

> We thus rather follow Nature, who producing nothing vain or superfluous often prefers to endow one cause with many effects. Though these views are difficult, contrary to expectations, and certainly unusual, yet in the sequel we shall, God willing, make them abundantly clear at least to mathematicians. (*De Revolutionibus.*)

For many years Copernicus had kept silence on his theory through fear, as he admitted, of ridicule; but when his great book was finally published he addressed it with reverence to the Pope himself. It was a time when reform of the calendar was being discussed in the Vatican and he offered his system as being a better basis for calculation than the established Ptolemaic one. The book was accepted by the Catholic authorities in the same spirit without thought of heresy and many years were to pass before it was placed on the notorious 'Index' of books which were not to be read by good Catholics. Luther poured scorn on the theory as did many others but meanwhile it was slowly and peacefully gaining some adherents in academic circles throughout Europe.

STARS WITHOUT END

The first extension of the new system of cosmology was in the realm of the fixed stars. Although Copernicus had retained the crystal sphere in which they were embedded it had been reduced to rest. If this were so there seemed little reason left for the existence of this invisible sphere at all. The stars themselves were of differing brilliance and could easily be imagined to lie at different distances from the Earth. All at once it became possible to imagine an infinite, unbounded Universe—the idea, which had been mooted before, was now intoxicating! By the end of the century the Copernican system had many followers in England. Thomas Digges translated the first book of *De Revolutionibus* in 1576 and added his own enthusiastic views on the infinite extension of the stars:

> . . . we may easily consider what little portion of God's frame our Elementary corruptible world is, but never sufficiently be able to admire the immensity of the Reste. Especially that fixed Orbe garnished with lights innumerable and reaching up in Spherical altitude without ende.

Digges' ideas were not totally new and were couched in reassuringly religious terms but those of Giordano Bruno, an apostate monk who had fled to England at this time, were far more radical and alarming. He held not only that the universe was limitless but that there was an infinity of *inhabited* worlds within it.

> There are innumerable suns, and an infinite number of earths revolve around those suns, just as the seven we can observe revolve around this sun which is close to us.

Bruno was a self-avowed pantheist and a heretic, so, when he rashly returned to Italy, he was seized by the Vatican and after eight years in prison was condemned to death by the Inquisition while still expounding his mystical views and was burnt at the stake in the year 1600. This was the point at which the Church began to realise the possible dangers of the Copernican system although it was never directly mentioned at Bruno's trial. In Italy, at least, disapproval was now in the air as Galileo was later to discover to his cost!

During the century that followed the publication of Copernicus' theory the situation changed dramatically. From being an obscure mathematical alternative to the established method of calculating celestial motion it became a possible explanation of nature with startling observations to lend it conviction. When John Donne wrote *Ignatius His Conclave* (1610) he had described Copernicus beating on the doors of Hell saying, 'Shall these gates be open to such as have innovated in small matters? And shall they be shut against me, who have turned the whole frame of the world, and am thereby almost a new Creator?' But Ignatius threw out his claim: '... Let therefore this little Mathematician, dread Emperor, withdraw himself to his own company.' The 'little Mathematician' was now becoming in reality a shaker of the Universe.

THE CRYSTAL SPHERES DISAPPEAR

In 1577 a bright comet appeared over Europe, and in Denmark there was a great astronomer ready to observe it—Tycho Brahe. Comets

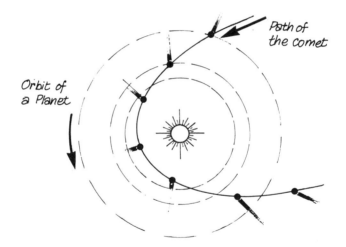

Orbit of a comet (probably part of a closed ellipse) compared with planetary orbits

of different degrees of brilliance appear from time to time in the night skies for all the world like some kind of celestial tadpole! They have at least one long tenuous tail which points away from the

36

Tycho Brahe's quadrant for fixing the position of the planets and comets

Sun whatever direction the comet is moving. After a few days or weeks they grow dim as they recede from our view again heading, it seems, right out of the solar system. Tycho Brahe traced out the movements of his comet with great care and was forced to the amazing conclusion that its elongated path cut right through the solid 'spheres' that were supposed to carry the planets. Without hesitation Tycho made the logical decision that the spheres, quite simply, did not exist. This made a vast difference to man's picture of the Universe. He wrote that

> . . . the machine of Heaven is not a hard and impervious body full of various real spheres, as up to now has been believed by most people. It will be proved that it extends everywhere, most fluid and simple, and nowhere presents obstacles as was formerly held, the circuits of the Planets being wholly free and without the labour and whirling round of any real spheres at all being divinely governed under a given law.

THE MIRACULOUS TELESCOPE

Tycho Brahe had no telescope but its discovery was not now far away, unfolding when it came, new and wonderful sights. In the hands of Galileo those 'marvels' were to shake the age-old system of Ptolemy to its very foundations. Of course the actual movement of the Earth could not be 'seen' directly but there were many observations that the established system could not begin to explain. Several conservative astronomers simply refused, ostrich-like, even to look through the notorious 'spy-glass'. Others, who did look, retorted that the new sights must be painted in some way on the inside of the instrument. Nothing dampened Galileo's jubilation and he published his first results in the famous *Starry Messanger* within a few months of making his telescope. Here are some of the more controversial results he obtained during his first few years of observation.

Replicas of two telescopes of Galileo made about 1610. The originals are preserved in Florence

(1) On the Moon there were clear and craggy mountains and large flat planes which he assumed to be seas. (We still call them 'maria' (singular, 'mare') to this day.) It followed that the Moon was not a perfect celestial orb as Aristotle and Plato had taught, but a rough, uneven body very like the Earth.

(2) Each of the planets was observably round and disc-like, although the stars, by contrast, appeared as distant points of light.

(3) Not only does the Earth have a Moon but so also does Jupiter— in fact, four moons—which could be clearly seen revolving round their parent planet.

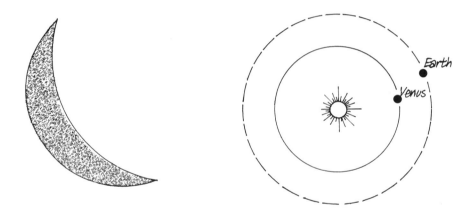

The 'Horned Venus' as seen from the Earth

(4) The planet Venus, whose orbit lies between us and the Sun, goes through phases like the Moon. This shows that it, too, like the Earth, is a non-luminous object shining only by reflected sunlight.

(5) There are dark spots on the surface of the Sun—another imperfection that Aristotle would never have approved—and observation of these showed that the Sun itself rotated on its axis about once in every month (27 days).

Although none of this constituted a proof of **the Copernican** system it did emphasise the similarity of the Earth to the six other planets then known. The Church, on the other hand, wanted to maintain its view of a central Earth—the only massive, corruptible, and inhabited body in the Universe. All other objects in the heavens were to be made of this remote celestial matter, aether, and to be beyond the realm of mortal man. The *De Revolutionibus* of Copernicus was finally placed on the Index in 1616, ominously soon

Galileo's sketch of the Moon (from Siderus Nuncius *1683)*

after the *Starry Messanger* was published. It was not the mathematics of Copernicus that were judged 'false and contrary to Holy Scripture' but the implied philosophy of the new system. Yet still Galileo wanted to marshal all the evidence and convince the world of the truth of his viewpoint. He waited until 1632, when he judged the new Pope favourable, to launch his great polemic, *Dialogue on the Great World Systems*. Throughout this long and lively argument he maintained, through the mouth of Salviati, not only the Copernican attitude, but also a belief that the Earth was not essentially different from the other planets. Another character, Simplicio, is allowed to present the traditional outlook.

> *Salviati.* . . . The Earth is deprived of light; the Sun is most splendid in itself and so are the fixed stars. The six planets do totally want light, as does the Earth; therefore their essence agrees with the Earth and differs from the Sun and stars. Therefore, the Earth is movable, and the Sun and starry sphere immovable.
>
> *Simplicio* . . . how much more proper a distribution, and more convenient with Nature, with God himself, the Architect, it is, to sequester the pure from the impure, the mortal from the immortal, as other Schools teach, which tells us that these impure and frail matters are contained within the narrow conclave of the lunar sphere, above which with uninterrupted series rise the things celestial. (*Dialogue on the Great World Systems.*)

But poor Simplicio was too often confused and confounded by Salviati and his friend so that, in the end, the *Dialogue* became more of a public vindication of Copernicus' and Galileo's beliefs than a true discussion. Galileo had gone too far! If nothing is special about the Earth, what then, thought the Cardinals, happens to the status of man and his relation to God?

The rest of the tragic story is well known. Galileo, now over seventy and ill, was summoned from Florence by the Holy Inquisition and travelled the two hundred miles painfully in a litter. Once in Rome he became a prisoner and, although the Copernican doctrine had never been officially pronounced a heresy, it was on this charge that he was interrogated under threat of torture and condemned. On 22 June 1633 the old man was dressed in the white shirt of penitence and made to kneel on the stone floor of the Dominican monastery while an account of 'errors' and the long sentence of the court was pronounced over him. Still on his knees, Galileo read the humble form of recantation of his heresies and of obedience to the Catholic Church. He was given a life-sentence which was later commuted to permanent house-arrest in his own farm near Florence. But in the end it was not the aged scientist who was defeated, for although a prisoner and nearly blind he lived to write another book full of more new ideas—it was Italian science itself that was destroyed. Galileo sent his last work to Leyden to be published and with it, to the north of Europe, went the living kernel of free scientific speculation.

4
Searching for a Force in Space

Galileo was a brilliant and famous figure in his day, a colourful and disputatious writer as well as a novel and practical experimenter. Through his work and championship the Copernican theory came to be widely known and discussed, but other urgent problems had been thrown up by this new doctrine which had already seized the imagination of one mystical German—Johannes Kepler—and were later to spill over into every centre of science. Kepler was a very different man from Galileo, ecstatic in his mathematics as well as in his religion and very few of his contemporaries understood, or even read, his books. Galileo wrote of him, 'I always value Kepler for his subtle free-ranging genius (perhaps too free) but my philosophy is very different from his.' Kepler corresponded with Galileo and sent him copies of his books but the latter rarely even answered his letters! In religion also the two men were poles apart. Despite persecution Galileo accepted his Catholicism uncritically; it was only in science that he demanded freedom of thought; but Kepler was a devout protestant and suffered an excommunication which distressed him over a difference of opinion about Lutheran doctrine that was quite unconnected with his scientific work. Like Newton after him, Kepler was deeply involved in astrology and biblical numerology—the date of Creation and the End of the World. Mathematical physics and occult religion may seem strange bedfellows but the vis on that inspired them both was of the greatest value to Science.

When Tycho Brahe threw away the solid crystalline spheres of the 'Heavenly Machine' he left the planets, moving distantly through space on their appointed courses to an incomprehensible

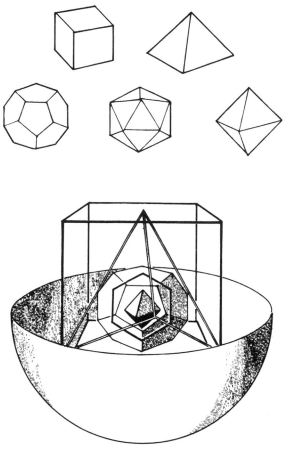

The perfect geometry of the planets, and the five regular solids that fit into spheres within each other giving the orbits (so Kepler hoped) of each of the planets, and a reason for the total number then known

rhythm as regular as the cycle of the seasons. No wonder that it was the mathematical and metaphysical Kepler who was drawn to this subject for he was ambitious to solve each numerical puzzle in the Universe. Geometry was, for him, part of the mind of God so that every number and every measurement in the heavens must fit into His great Scheme. Kepler wrote that he was looking for the 'wonderful correspondence of things' in the sky and in the harmony of music just as that other great mystical mathematician, Pythagoras, had done twenty-one centuries earlier.

KEPLER'S LAWS

As Tycho's heir, Kepler had access to all his careful observations and during the course of his life he derived from them three famous laws of planetary motion which bear his name to this day:

First Law
The planets all revolve in ellipses—not circles—with the Sun at one focus.

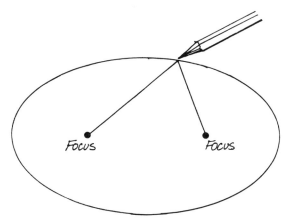

An ellipse can be drawn by placing the point of a pencil in a tight loop of string fixed between two points—the foci

Second Law
The planets all sweep out equal areas referred to the Sun, in equal times.

46

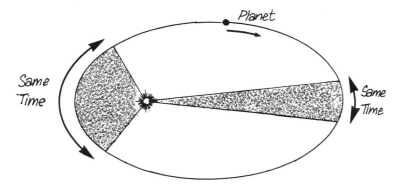

It follows that they travel slower in that part of their orbit which is farthest from the Sun

Third Law

The square of the total time of revolution of a planet is proportional to the cube of its distance from the Sun. (Therefore the more distant planets move slower in their orbits than do the nearer ones.)

The second and third laws that Kepler formulated showed clearly that the farther a planet was from the Sun the less was the force which moved it in its slow impressive measure through the sky. He felt sure that the source of this motion was the Sun itself but, understanding little of centrifugal force, he could not reconcile the focal position of the Sun with the elliptical movement of the planets. His first idea was that the driving power of the Sun had to be like that of a great eccentric whirlpool.

> The sun remains in its place, but it rotates as if upon a lathe and sends out from itself into the depths of the Universe an immaterial species of its body analogous to the immaterial species of its light. This species turns with the rotation of the Sun after the manner of a most rapid whirlpool throughout the whole extent of the Universe, and it bears the planets along with it in a circle with a stronger or weaker thrust according as, by the law of its emanation, it is denser or rarer.

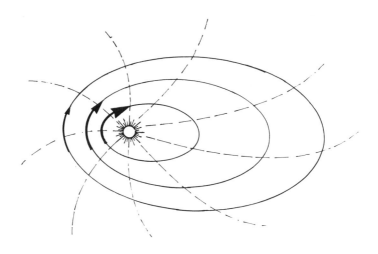

The 'Cosmic Whirlpool' (from the New Astronomy)

When Galileo's telescopic discoveries later showed that the Sun does actually rotate Kepler was jubilant. He himself had only a poor telescope and was hampered by bad eyesight but when Galileo sent him a copy of his *Starry Messanger* he immediately published an enthusiastic commentary on it. One point that Kepler made at this time shows just how far he had thrown off medieval ideas and was heralding the coming of empty Newtonian space.

> Under your guidance I recognise that the celestial substance is incredibly tenuous. . . . A single fragment of the lens interposes much more matter (opacity) between the eye and the object viewed than does the entire vast region of the aether. . . . Hence we must virtually concede, it seems, that the whole immense space is a vacuum.

Kepler continued to speculate throughout his life on the nature of the power which keeps the planets in their elliptical orbits

48

through pathless space. He had studied and written on the science of
optics and had derived a law showing how illumination decreased
with the square of the distance from a bright source just as the
force on the planets seemed to do.

Still influenced by the old Christian reverence for light he would
probably have liked to explain the Sun's force in terms of its
'immaterial species'. But had light really sufficient mechanical thrust
to move the heavy planets, and, if so, why did they not come to
rest during an eclipse of the Sun?

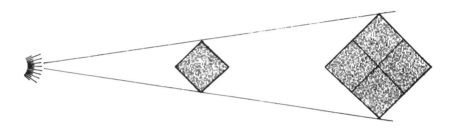

*Twice the distance from the source—therefore one-quarter
the illumination*

Massive objects are obviously pulled strongly by gravity but this
had always been considered as an Earth-bound force. Now that the
Earth was no longer at the centre of the Universe perhaps gravity
too could be viewed on a cosmic level? Kepler was ready to assert
that any two bodies will attract each other and move together; he
was even prepared to extend the power of gravity from the Moon
to the Earth in order to account for the monthly variation in the
tides. But then the Earth and the Moon could be seen to have a lot
in common—mountains, seas, and the same length of yearly revolution round the Sun. Gravity might affect them both and yet not, in
Kepler's eyes, pervade the whole solar system.

49

The force between the Sun and its planets was held to be less 'mundane', indeed Kepler considered very carefully the medieval idea that the planets might be moved by 'soul' or by 'Intelligences' within them like the angels who turned the crystal spheres. He was even sufficiently influenced by bygone theory to hold that such divine 'Intelligence' would be sure to cause strictly circular motion and, as he had already computed the orbit of Mars as an ellipse, he was forced to dispense with the services of such perfect beings! Yet still the problem of the planets' movement in their remote courses remained to be solved. If the force which propelled them was neither Soul, nor light, nor terrestial gravity, could it be due to the mysterious power of Magnetism?

THE FIRST STUDY OF MAGNETISM

In 1600 a famous book on Magnetism had been published in England by Queen Elizabeth's own physician, Dr William Gilbert. It was a fascinating study full of experimental work on both magnetism and static electricity, and copiously illustrated. As was usual at that time, it was written in Latin and was widely read throughout Europe. Galileo himself rashly quoted Gilbert's work with praise in his ill-fated *Dialogue on the Great World Systems* and the fact was noted at his trial when Gilbert was described as 'a perverse heretic and a cavillous and contentious defender of Copernicus'. Kepler, too, had read this book and admired its conclusions. Magnetism had waited long for an exponent and now, when astronomers were casting around for a force that could influence the planets from a great distance, the whole subject became alive with cosmic significance.

Both the magnetism of a lodestone and the electricity which is excited by rubbing amber can attract substances; but Gilbert seems to have been the first to distinguish carefully between them. He made himself a kind of primitive electroscope—a light metal needle pivoted on a sharp point—to examine objects which had been charged by rubbing. He found that not only amber but also sealing-wax, jet, glass, sulphur, and a multitude of other substances could pull the pointer round. He showed experimentally that no air currents were responsible for this force (as Plato had postulated in his theory of 'circular thrust') and that the electric and magnetic 'emanations' responsible for attraction were, in the language of the age, very 'subtle', far rarer than air itself. This was a step forward towards the idea of completely immaterial influence through space.

Gilbert's 'versorum' attracted to rubbed amber

Gilbert is most celebrated for the 'model Earth' that he made out of a lodestone to demonstrate the magnetic properties of the world. It successfully duplicated all the terrestial behaviour of compasses and won him such fame that it has become an almost invariable symbol in his portraits. The diagram illustrated shows the clear sphere of influence of the lodestone in which its force is exerted something like the modern concept of a 'magnetic field'.

Gilbert was a passionate Copernican and was anxious to use his knowledge of magnetism on a vaster scale to explain the heavenly motions. He could show in his own laboratory that a floating lodestone would rotate so that its north–south axis pointed to the north of the world and it seemed a great triumph to him. It proved that an

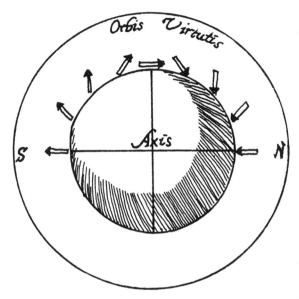

The 'Terrella' (showing the 'dip' of a compass on its surface)
Almost an exact copy of Gilbert's own illustration

earthly object could rotate naturally as Copernicus had taught and did not just move in straight lines up and down as the Aristotelians still held. From this experiment also he argued that the Earth's axis is orientated in space by magnetic attraction towards the Pole Star (the Cynosure or 'dog's tail').

> The whole earth would act in the same way, were the north pole turned aside from its true direction; for that pole would go back, in circular motion of the whole, towards Cynosura.

This hypothesis did not explain the rotation of the Earth, as some seemed to imagine, and indeed it turned out to be untenable in every way, but it paved the way for other speculations on the nature of forces in space.

Characteristically Kepler was eager to seize on this idea of mag-

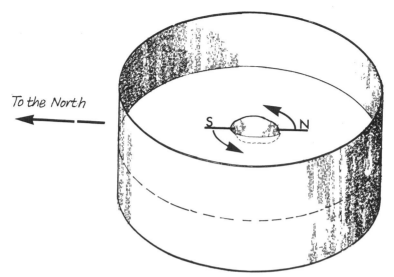

The floating 'terrella' spins round to point north

netic influence in space and he used it to explain the alternate speeding up and slowing down of the planets as they move round the sun in their elliptical orbits. Although this use of magnetism in the heavens proved, like Gilbert's own theory, to be quite unconvincing, it did successfully by-pass the intermediate aether and

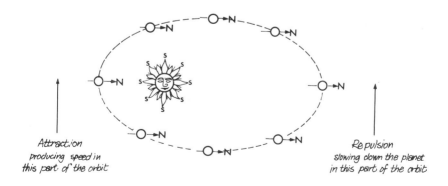

Attraction
producing speed in
this part of the orbit

Repulsion
slowing down the planet
in this part of the orbit

considered only the actual attraction of the Sun—unlike Kepler's earlier whirlpool idea.

The physicists of the next generation had to face this very choice—whether to summon the aid of a 'subtil aether' to provide the whirling mechanism, or to concentrate on a Sun-based force acting in some unspecified and mysterious way on the distant planets.

5
Gravity at Last

By this time the whole fabric of Aristotelean philosophy seemed to be full of holes. This view of the world was still taught in schools and colleges but its authority was crumbling. Many educated men read the works of Gilbert and Galileo; the poet John Donne even struggled through some of the more abstruse books of Kepler, writing sadly 'the reason which moved Aristotle seems now to be utterly defeated'. The second half of the seventeenth century was to see the excited exploitation of this scientific freedom and the founding of new scientific societies, but meanwhile the poet mourned the passing of a secure philosophy that had survived more than a thousand years. No doubt many shared his feelings of loss and dread.

> And new Philosophy calls all in doubt,
> The Element of fire is quite put out;
> The Sun is lost, and th' earth, and no man's wit
> Can well direct him where to looke for it.
> And freely men confesse that this world's spent,
> When in the Planets, and the Firmament
> They seek so many new; then see that this
> Is crumbled out againe to his Atomies.
> 'Tis all in peeces, all cohaerence gone;
> All just supply, and all Relation; . . .
>
> This is the world's condition now, and now
> She that should all parts to reunion bow
> She that had all Magnetique force alone
> To draw, and fasten sundred parts in one;

Shee, shee is dead; shee's dead : when thou knowst this,
Thou knowst how lame a cripple this world is.
(from *An Anatomie of the World*, John Donne.)

Two attempts to replace the old Aristotelean science by another complete system were made—by Francis Bacon in England and by René Descartes in France; but new 'rules' for investigating the world were not what the age required. Once the old shackles were broken the way had to be kept clear for the free play of scientific intuition, new mathematical methods, and the imaginative use of experiment. Somehow the more scientists peered over their own shoulders at their methods of working, the less they seemed to achieve! Francis Bacon's experimental philosophy contributed no fruitful hypotheses to science and though Descartes' logical reasoning produced theories that were grand and universal enough they proved untenable in the face of experimental trial. In the end it was the intuitive and mathematical theory of Isaac Newton, which was designed to fit the known laws and to predict new ones, that became the crowning achievement of the century.

By this time it was clear that the questions raised by the mechanics of the Solar System fell into three main categories :

(a) What is the nature of the *force* between Sun and planet, between planet and moon ?

(b) What is the nature and function (if any) of the *space* between them ?

(c) Can the known *movements* of the planets be derived from this force ?

NEWTON'S LAWS OF MOTION

Descartes was pre-eminently a philosopher and tried to answer the first two questions but failed on the third—the vital experimental testing ground. Isaac Newton, however, ignored the first question completely, stating firmly 'I frame no hypotheses'. He attempted to answer the second which was useful and realistic (if irritatingly illogical to the French school!) and won, on the third, a resounding triumph which established his theory, in the teeth of all philosophical criticism for two busy centuries.

It seems clear to us now, with the easy wisdom of hindsight, that the reason for Kepler's final failure to identify the force that spins the planets round the Sun, was a basic lack of understanding of the very nature of force itself. Since the time of Galileo scientists

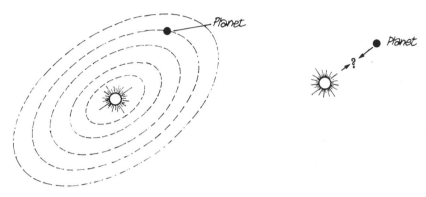

The vortex of Descartes and the mutual reaction of Newton

had been groping their way towards clear definitions of the concepts of 'acceleration', 'fall', 'inertia', and 'force' and some progress had been achieved. The basic dilemma had been a legacy of Aristotelean thought whereby motion beyond the earth was held to be *naturally* circular. On the earth an object can only be spun in a circle by a force towards the centre, as a discus is whirled round by the strength in the athlete's arm. When he leaves go the discus will 'fly off at a tangent' in a straight line. Which movement is natural and unforced, the *circle* or the *straight line*? Newton had no doubt that it was steady speed in a straight line which represented the free 'coasting' of any uninhibited body anywhere in space, and that change of direction, as much as slowing down or speeding up, in-

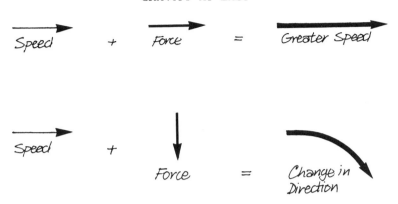

dicated that an external force was acting on the object. On these principles the approximately circular movement of the planets implied a continual change in direction, and therefore it needed a steady force towards the centre of the circle. The pull of the Sun must cause the planets to swerve inwards (instead of careering off along the tangent) so that they were perpetually 'falling' towards the Sun although, as a result, they continued to revolve steadily about it in a circular path at a constant distance! Once this paradox

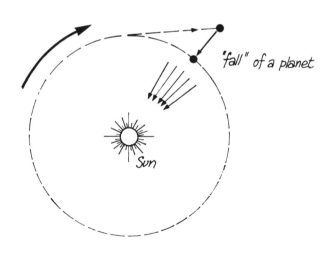

was cleared up Newton could calculate the magnitude of the centrifugal force acting on the planets in terms of their masses, their speeds, and the radii of their orbits round the Sun. This was to prove the vital 'missing link' in planetary mechanics.

GRAVITY—FROM AN APPLE TO THE MOON

In the years 1655–6 the colleges at Cambridge University were closed down because of an outbreak of plague and the young Newton, who was then in his twenty-fourth year, returned home and worked out a new branch of mathematics, which he used to develop a new science of mechanics, which supplied a new theory of gravitation—as well as some other startling achievements! Newton had barely ceased to be a student and yet, alone on his farm, and still in his early twenties, he outlined in those two active years the creative output of a lifetime.

Newton had already studied Kepler's work and had managed to extract the three laws of planetary motion from the mass of mystical

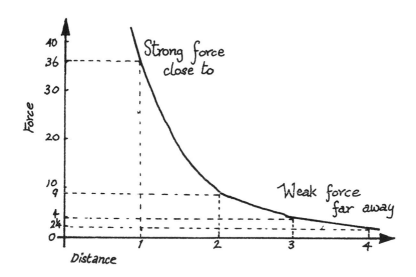

speculation. He then combined them with his newly found formula for the central force on a revolving body and deduced that the cosmic power he sought should obey an Inverse Square Law—like the law for light illumination that Kepler himself had discovered. (Arithmetically this means that if the distance from the attracting object is *doubled* the force is reduced to a *quarter*, if the distance is multiplied by *three* the force is reduced to *one-ninth*.)

Newton could show that the force from a magnet obeyed no such law and even Kepler had known that light, although it obeyed the right law, could not be the agency of celestial movement. The unknown force was clearly directed towards the centre of a massive body (sun or planet) and, to Newton, this suggested common gravity—the cause of weight and fall on earth. So, using his new 'Inverse Square' law he set out to identify the distant force which makes the Moon 'fall' towards the Earth in her orbit with that which makes an apple fall from an earthly tree. The numbers he first used were not very accurate but he wrote, years later, that he had compared his calculated value with the established measurements and 'found they answered pretty nearly'. The following passage tells, in his own words, how Newton fitted the pull of ordinary gravity to the rôle of moon-and-planet-mover and this great feat has remained unchallenged to the present day.

> The mean distance of the moon from the earth . . . [is about] 60 diameters [of the earth]. . . . And now if we imagine the moon, deprived of all motion to be let go, so as to descend towards the earth with the impulse of all that force by which it is retained in its orb, it will in the space of one minute of time describe in its fall $15\frac{1}{2}$ Paris feet. . . . Wherefore, since that force, in approaching the earth, increases in the proportion of the inverse square of the distance, and, upon that account on the surface of the earth is (60×60) times greater than at the moon, a body in our regions falling with that force, ought in the space of one minute of time, to describe $60 \times 60 \times 15\frac{1}{2}$ Paris feet. . . . and with this very force we actually find that bodies here upon earth do really descend. . . .

And therefore the force by which the Moon is retained in its orbit is that very same force which we commonly call gravity. (Newton, *Principia*.)

Newton's calculation did not have to be repeated with objects of different weight, nor did he need to know the mass of the moon for it had long ago been established that all objects have the same acceleration as they fall to the ground. This is a characteristic of gravity. Galileo is popularly supposed—probably quite untruly—

Vacuum

to have dropped cannon balls of different weights from the Leaning Tower of Pisa to verify this point. More recently Robert Boyle had used his new vacuum pump to compare the fall of a guinea and a feather under conditions where there could be no air resistance. With startling rapidity the feather plummeted down the evacuated tube, side by side with the golden guinea! Newton performed some experiments of his own on the same subject designed to show up any smaller discrepancies. He made a pendulum with a hollow bob and simply timed its period of swing when filled with different weights. He could detect no differences at all. It was clear from these experiments that every object, heavy or light, undergoes the same acceleration in the gravitational pull of the earth. But since a heavy object needs a greater force to get it moving than does a light one it must follow that the force of gravity on any object is exactly proportional to the mass of that object—large for a heavy body, small for a light one. (This, as Newton pointed out, is another feature that distinguishes gravity from magnetism.) Finally, since there are two objects involved in gravitational attraction and the pull between them is mutual, Newton's law of Universal Gravitation took the form

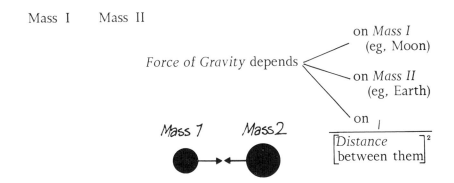

It was not until 1686 that Newton published his theory in *Principia Mathematica* of which Book III, 'The System of the World', contains most of his work on gravitation. In the intervening twenty years Newton had become a Cambridge Professor and an illustrious member of the new Royal Society, known alike for his touchy pride and brilliant mathematics. It was only in 1685, at the personal request of Edmund Halley and Christopher Wren, that he made public his derivation of the Inverse Square Law from the difficult elliptical orbits of the planets and began to prepare his great book for publication.

VACUUM OR AETHER?

Newton was very concerned that his work should present as logical and complete an argument as possible, so he set out his views on both space and time before proceeding to the host of proofs and propositions on motion that he needed to place his Law of Gravitation on a firm mathematical basis. This is the part of his work that has not stood the test of time so well and yet it, too, is fascinating because it contains all the challenge that Relativity had to take up and answer with such exciting consequences in the present century. But in his own age Newton's views were almost equally revolutionary and met with considerable opposition.

To Descartes and his followers on the continent the age-old aether was still a very real substance pervading the whole of space. Its vortices spun the planets round in their orbits and it was the vehicle of light itself. Descartes had based his influential philosophy on the maxim *'cogito, ergo sum'*—and so, if by taking thought he could verify his own existence, it followed that anything which was unthinkable, like the 'nothingness' of space, simply could not exist! The vacuum in a barometer tube might be empty of air but it was still quite full of the elusive aether; indeed Descartes held that a box containing *nothing* would collapse instantly—not because of atmospheric pressure acting on it from the outside, but because a total void was unimaginable and so non-existent. This belief in continuous aether throughout what we call space endowed every point in it with measurable position and movement.

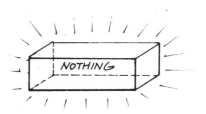

The hollow box with an unimaginable interior

For the young Newton, on the other hand, the aether had no part to play in celestial mechanics, and in the first few paragraphs of the *Principia* he wrote: 'I have no regard in this place to a medium, if any such there is, that freely pervades the interstices between the parts of bodies.' He was bold enough to imagine motion 'in an immense vacuum, where there was nothing external or sensible with which the globes could be compared'. Did he then conceive of space as infinite, without any point of reference and all movements in it as merely relative? It seems that he forced himself to the brink of this stupendous modern viewpoint only to draw back from its immensity. Later, in Book III of the *Principia* he presents simply and comfortably as his first Great Hypothesis:

> *That the centre of the system of the world is immovable.* This is acknowledged by all while some contend that the earth, others that the sun, is fixed at that centre. Let us see what may from hence follow.

This apparent naïvety is based on a famous thought experiment. If a pail of water were suspended in space it would be quite possible to distinguish between the two cases:

(*a*) when the pail was at rest and the stars moved around, or (*b*)

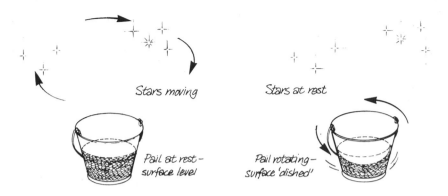

when the pail was rotating while the stars were at rest. In case (a) the surface of the water in the bucket would be flat and in case (b) the surface of the water would be curved. Since this provided an observable test of absolute motion in space there must be some fixed point in the Universe to which this movement could be referred. Where this focal point of all celestial movement was located was unknown—except perhaps to the Creator—but it did make empty Space itself 'absolute' and full of mathematical points each one of which could bear its own clear 'milestone' from the centre.

About the nature of time, Newton had no doubt at all. He wrote, 'Absolute, true and mathematical time, of itself, and from its own nature, flows equally without relation to anything external, and by another name is called duration.

FAME AND CONTROVERSY

The success of Newton's theory of Gravitation was hailed jubilantly in England and celebrated in verse both long and short; memorably by Pope's famous 'epitaph':

> Nature and Nature's Laws lay hid in Night,
> God said 'Let Newton be'—and all was Light.

The enthusiasm was distinctly less on the continent. Newton himself realised quite clearly the size of the questions that he had *not* answered. What was gravity? How could it move distant objects without contact? He insisted that 'action at a distance' was clearly absurd but was chary at first of making any guesses as to the mechanism by which it might work. His admiring English followers, however, saw no faults in the theory and championed it uncritically.

By the end of the seventeenth century European science had a very different complexion from that which it had worn before. Gone were the days when books were written in Latin—the universal language of scholarship—and when scientists, like Galileo and Kepler, communicated to each other freely across all national boundaries. Science was now seen as an added lustre to the glory of a particular country, almost a measurable political asset! Louis XIV was advised by his Prime Minister to inaugurate a French Acadamie des Sciences to contribute to the magnificence of Versailles, and to the brilliance of his regime. Charles II of England had less wealth but he too was keen to advance the nation's prestige and he granted the Royal Society its charter soon after the Restoration. Both these innovations—writing in the vernacular and royal patronage—might well be expected to make scientists more insular and patriotic; in fact violent controversies did rage across the Channel on at least three separate counts by the turn of the century, and all of these were centred on the irascible and autocratic figure of Sir Isaac Newton. The followers of Descartes poured scorn on his gravitational theory, maintaining that only 'scholastic, occult qualities' could act at a distance through space and the careful, logical Newton writhed in fury at the charge! There was also an argument with Christiaan Huygens about the nature of light in which the English suffered by their ardent championship of Newton, and lastly a bitter running battle ensued with Leibniz as to who first invented the calculus. It was a hot-blooded age and quarrels over

An engraving made to celebrate the founding of the Royal Society.
It was a conscious effort to glorify one nation's science

philosophy and mathematics were carried on with the same venom as were those in religion and politics. Epithets like 'pig', 'dog', and 'rogue' were commonly hurled at one's opponents and august members of the Royal Society were even known to pull faces at each other during proceedings! Princess Caroline, who corresponded regularly with Leibniz on scientific matters, wrote to him about his incessant quarrel with Newton: 'But great men are like women, who never give up their lovers except with the utmost chagrin and mortal anger.'

MORE 'INVERSE SQUARE' FORCES

Eventually Newton's concept of Universal Gravitation was accepted throughout Europe on the recommendation of its superb mathemati-

cal solution of the planets' movements. Even some philosophers, like Voltaire and Kant, were eager to lend their support in spite of the difficulties of understanding how one remote massive object could possibly pull at another without any contact between them. Once this unlikely notion was swallowed the Inverse Square Law could be seen as an inevitable result for any influence which could spread out through a neutral space. Such laws were known to exist also for light and sound just because space is three-dimensional—

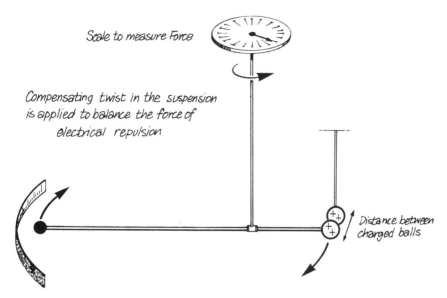

Scale to measure Force

Compensating twist in the suspension is applied to balance the force of electrical repulsion

Distance between charged balls

Coulomb's Experimental Proof of the Inverse Square Law for Electric Repulsion (1785)

so that an area more distant from the source *must* receive less effect in this way. Then, during the eighteenth century the forces of attraction and repulsion produced by magnets and electric charges were carefully studied and measured. It was soon established that they, too, like the force of gravity, could act through a vacuum and would obey the same inevitable Inverse Square Law. The old theories of mysterious 'emanations' were at last laid to rest and the odd idea of 'action at a distance' became quite respectable. It seemed that a beautiful simplicity reigned in Physics by which *all* the remote influences had at last been tamed to mathematical conformity and that the intervening space could be comfortably ignored.

6
The Great Debate on Light

Yet the aether never died. When Newton was an older man his inclination turned away from cautious mathematical reserve and he began to speculate more freely on chemistry, theology, and even astrology. In his second great scientific book *The Opticks*, he denied categorically that the cause of gravity was in the mass of an object but visualised it prophetically, as a property of space itself—of that great sea of aether which fills the Universe. The seventeenth century had also seen the birth of a new study of light and every theory—including Newton's own—had needed the services of an intangible, elastic fluid to speed the rays on their mysterious way through space.

> . . . if the elastic force of this [Aetherial] Medium be exceeding great, it may suffice to impel Bodies from the denser parts of the Medium towards the rarer, with all that power which we call Gravity. (*The Opticks*, 1704.)

At long last the study of light had left the realms of Metaphysics and become a part of the new Natural Philosophy. The 'souls', 'forms', or 'simulacrums' which the schoolmen of the Middle Ages had held to be impressed through an invisible medium, gave way to more useful work on lenses and prisms, on the laws of angular deviation, and of the composition of light itself. The spur to this new approach seems to have been Galileo's controversial telescope. Lenses had been ground by artisans for nearly three centuries before this but it had needed the wonderful views of mountains on the Moon and of the satellites of Jupiter to shake scientists out of their indifference. Sunlight was known to bend as it entered a transparent substance ('pellucid' was their word for it) and to produce a vivid spectrum of colours as it passed through a glass prism and yet no one was sure if light actually took time to travel—as sound clearly did—nor even if it moved from the object to the eye or vice versa! Galileo, in his forthright way, had tried to measure the speed of light by using two observers equipped with lanterns on neighbouring hill-tops, but its speed proved too great to detect. Eventually it was the moons of Jupiter, which he had been the first man to see, that helped solve this fundamental point.

THE SPEED OF LIGHT

The moon closest to Jupiter takes less than two days to complete one circuit and re-emerge from behind its parent planet. Meanwhile the Earth, of course, is moving on its regular path around the Sun. Copious observations had been made on the times of eclipses of this tiny moon and it was shown that when the Earth was receding from Jupiter in its orbit the intervals between the reappearances of the satellite were slightly too long, and six months later when the Earth was approaching Jupiter they were slightly too short. This clue was enough; it had to be the time taken by the light itself as it 'caught up' with the retreating Earth that caused the anomaly. In 1666 a Danish astronomer, Ole Roemer, working at the Paris Observatory, made the necessary calculations and predicted that the next emergence of Jupiter's moon would be ten minutes later than the time calculated from the August observations. He made a careful check on 9 November and was proved triumphantly correct. His value for the speed of light gave the astonishing result that it would take light a bare 11 minutes to travel the 93 million miles from the Sun to the Earth. Even this fantastic speed proved to be an underestimation. The true figure should be nearer 8 minutes—equivalent to about 300,000,000 metres or 186,000 miles each *second*!

Knowing that the velocity of light, although enormously large, was not infinite was a great stimulus to new speculation in the latter

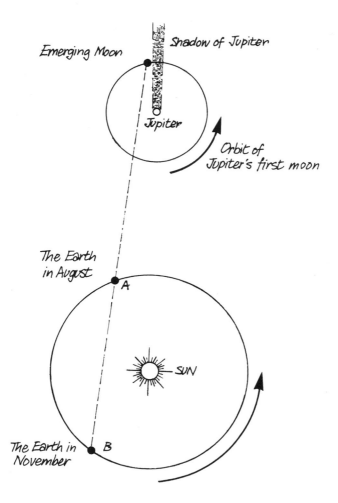

Roemer's Measurement of the Speed of Light. *AB is the extra distance light from the Moon must travel to 'catch' the Earth! (during 40 circuits about Jupiter)*

half of the century. In what way did light speed through space? And—in effect much the same point—what is light? Two famous scientists, Christiaan Huygens and Isaac Newton, attempted answers to these very basic questions and, in the careful spirit of the 'New Philosophy', tested their assumptions against every known phenomenon of light.

THE 'CORPUSCULAR' THEORY

Almost by instinct Newton chose to regard light as a stream of swift particles. He had already worked out the laws of motion for ordinary particles almost single-handed and he manipulated them with great skill. The most obvious property of light was its tendency

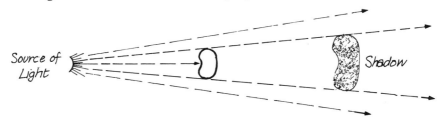

'Tracks' of Newton's tiny particles of light

to travel in straight lines like an ideal projectile in empty space and his mechanical laws for impact and recoil could be easily adapted to explain the reflection of the light corpuscles on any polished surface. Obviously Newton could never *prove* his hypothesis but the form in which he phrased his classic 'Query' leaves no doubt as to his own conviction:

> Are not the Rays of Light very small Bodies emitted from shining Substances? For such Bodies will pass through uniform Mediums in right Lines without bending into the Shadow, which is the Nature of the Rays of Light. . . . (*Opticks*.)

The refraction of light—the way in which oblique rays of light bend when entering any transparent substance—could be due to attraction of the particles, another concept which was familiar to him from gravitational theory. He found that this idea too gave rise to the correct Law of Refraction. His experiments with glass prisms are justly famous for he was the first to show, by a whole series of careful observations, that white light was really a mixture

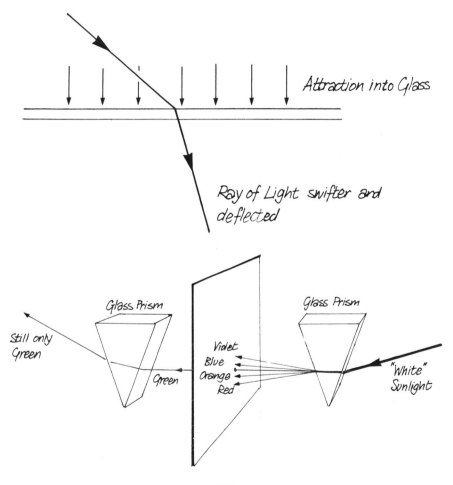

of many colours. Some had held that the glass prism only spun the particles of light at different speed and that this gave them different colours, but Newton proved that the prism merely deflected the different colours of which sunlight is composed by different amounts and that this only separated the already existing colours. Then he showed that by recombining the spectrum he could again build up the colour we call white.

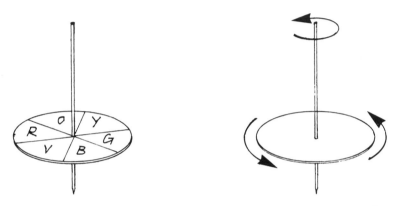

Disc painted with the colours of the rainbow looks white when spinning quickly

Yet the 'Corpuscular Hypothesis' itself did meet with some serious problems. How could the surface of a glass block both reflect (repel the particle) and refract (attract it) at the same time? How could a very narrow slit actually cause the light to spread out and give an image with coloured fringes? In the face of these difficulties Newton imagined that his 'swift bodies' generated tiny waves which spread out ahead of them through the invisible aether like ripples on the surface of a pond and set the particles themselves vibrating into fits of easy reflection and easy transmission. His estimation of the time intervals between these alternate 'fits' differed for each colour so as to explain the iridescent effects seen on soap bubbles and on thin

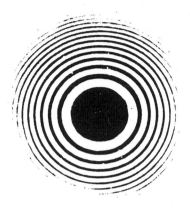

Newton's Rings

films of oil. He also applied this new idea to the beautiful series of concentric rings visible in the air space between a lens and a flat plate of glass.

Now this was no longer a strictly Particle theory of the nature of light and although Newton's reputation ensured that his original idea would have massive support, he himself seemed to be moving towards this different viewpoint.

> . . . when a Ray of Light falls upon the Surface of any pellucid Body . . . may not Waves of Vibrations, or Tremors, be thereby excited in the refracting or reflecting Medium. . . ? And do they not overtake the Rays of Light, and by overtaking them . . . they may be alternately accelerated and retarded by the Vibrations. . . ? (*Opticks.*)

Taken altogether Newton's theory was complex and not always self-consistent. Each new group of effects seemed to require new modifications to the simple original thesis of 'very small bodies'.

77

When it was also shown that asymmetrical crystals could split a single ray of light into two Newton had to tailor his luminous particles accordingly and give them an inherent 'lop-sidedness' in the form of two flattened sides so that they could lock into the crystal planes in two different directions. It is always hoped that

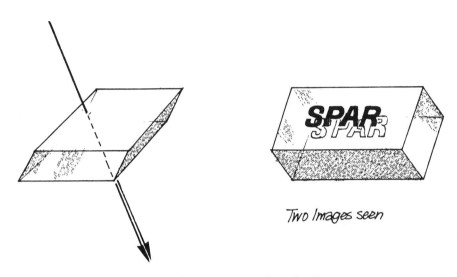

Two Images seen

Double Refraction through Iceland Spar

one inspired scientific idea will explain many different results with a simple inevitability which is totally convincing. Newton's basic hypothesis on the nature of light did not really pass this test although all his impressive and beautiful experiments with prisms did show beyond doubt that plain white sunlight was already an amalgam of all the brilliant colours of the rainbow. This point alone was quite difficult enough for his contemporaries to grasp. His other experiments on colour had to wait for a convincing explanation until a much more complete Wave Theory of Light was evolved in the nineteenth century.

78

THE WAVE-THEORY

There was also a contemporary Wave Theory of Light due to Christiaan Huygens, a Dutch mathematician living in France which, on balance, was just as successful as Newton's Corpuscular theory although it started from such different premises. Huygens justified his hypothesis on the grounds that it was only hot objects, fire and flame-containing particles in rapid motion, that give out light—just as the vibration of a violin string gives out sound. He also pointed out that light rays can cross right through each other without hindrance as can water waves and ripples. How then, he asked, could light consist of a stream of hard material objects?

Waves are something quite different. There must always be a material present—particles, fluid, jelly, or solid—but it is not the medium itself that actually travels, only a disturbance that runs through it like a whisper through a crowd. Each particle is jolted and, in turn, jolts its neighbour so that, like running a finger down a comb, it seems as if the jolt itself is moving although we know that each particle returns to its undisturbed position as soon as the

commotion passes it. Huygens believed, with Descartes, that the aether was composed of myriads of tiny particles and he visualised a percussion wave travelling through them at great speed like the impulse through a row of marbles. There was nothing occult about

Impinging marble causes a shock-wave to travel through the row and to throw off the end marble

this theory, as it was as mathematical and mechanical a model as Newton's own. 'This', wrote Huygens, 'is assuredly the mark of motion at least in the true Philosophy, in which one conceives the causes of all natural effects in terms of mechanical motions.' In space one particle of aether could impinge on several others and send out its effects in all directions like spherical ripples. On this analogy he based his theory of light waves. He showed how these spherical 'wavelets' may be constructed, how they superimpose and give rise to new wave fronts. It was an elegant theory, like a kind of

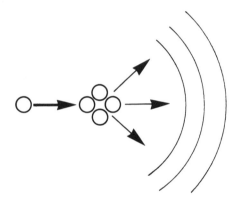

Spherical 'ripples' spread out from a multiple impact

moving geometry in space, and Huygens could prove that all the known laws of reflection and refraction arise quite naturally from its spreading surfaces.

These two theories of light then confronted each other without any decisive outcome for more than a century. Light was the only probe then known which could traverse the great spaces between the stars and these theories of light show very clearly the two contrasting views then held about space. In England, under Newton's influence, space was nothing but an immense vacuum in which gravity acted mysteriously at-a-distance and so it needed real light 'bullets' to pass through it. On the continent, under Descartes' influence, space was seen as an endless ocean of aether through which light could spread like shock-waves through the air. Victory for either theory might be expected to solve the riddle of space in a direct and simple fashion.

7
Philosophical Speculation on Space and Time

What happened in the eighteenth century? The seventeenth century had been so fertile in great men of science from Galileo and Kepler to Newton and Huygens. Enormous strides forward had been made in physics and were to be made again in the nineteenth century; why, then, were the intervening hundred years so barren? It is true that experimental work continued and considerable progress was made at this level but one looks in vain for a scientist of real stature, for a man of great ideas. There were many gifted amateurs —clergymen, like Priestley, statesmen like Franklin, or aristocrats like Cavendish, yet they give an impression of dilettantism when compared with the excited intellectual involvement of their predecessors. Maybe this dearth within science reflects a certain lack of genius in the whole realm of imaginative endeavour throughout this century. Neither in art nor in poetry are the greatest heights reached but, though this may provide an intriguing field for speculation, in physics specifically some possible reasons for this sterility can be seen quite early in the period.

ATTACK ON NEWTON'S CALCULUS

No sooner was the stranglehold of ancient philosophy broken than new schools of European philosophy began to grow up. The seventeenth century marks the turning point between these two trends of thought. When Newton did his most creative work he was free to settle his own problems of space and motion, of experiment and mathematics. There were so many different rôles he could play— a one-man-band with no conductor! But the philosophers of the eighteenth century applied logical tests to both mathematics and physics, defining their terms and criticising their methods while paying lip-service to their achievements. In 1734, for example, Bishop Berkeley published a powerful tract against the basic concepts of Newtonian Calculus and concluded with a series of pointed 'queries' in conscious parody of Newton's own style.

> Qu. 4. Whether men may properly be said to proceed in a scientific method, without clearly conceiving the object they are conversant about. . . ?
> Qu. 8. Whether the notions of absolute time, absolute place and absolute motion be not most abstractly metaphysical. . . ?
> Qu. 64. Whether mathematicians as cry out against mysteries have ever examined their own principles. . . ? (*The Analyst*. A discourse addressed to an infidel mathematician.)

IMPONDERABLE FLUIDS

In fact the formulation of Calculus was gradually improved and it ceased to operate on 'the ghosts of departed quantities', as Berkeley had called infinitesimals, and became instead the mathematics of smooth limiting values. This was the line of approach that led Leonhard Euler, the greatest mathematician of the age, to evolve equations of motion for flowing, swirling liquids just as Newton had done for solid moving objects. Fluids were very popular scientific models throughout the eighteenth century and were used to describe the 'flow' of electricity, of heat, and even of magnetism. Perhaps this was because both heat and electric charge seemed able to move without loss from one body to another, 'filling' them to different 'levels' of temperature or potential as they penetrated into the pores of the material.

These fluids—weightless, elastic, and self-repelling—do not seem very credible models to us now, hardly recognisable, in fact, as real matter at all.

This unmechanical attitude in science may also have been a reflection of the contemporary philosophy. Berkeley, Locke, and Hume had applied such stringent logical tests to the whole question of perception and reality that they were left with belief only in the validity of their own sense perceptions. The objects themselves, they held, could not be said to 'exist' in any way above and beyond our sensing of them. This attitude—sometimes tempered by religion—is well summed up in these two famous limericks.

There was a young man who said, 'God
Must think it exceedingly odd
If he finds that this tree
Continues to be
When there's no one about in the Quad.'

Dear sir:
Your astonishment's odd:
I am always about in the Quad.
And that's why the tree
Will continue to be,
Since observed by
Yours faithfully,
God.

So to this English school of philosophy there was no *real* material and this was hardly a conclusion favourable to the growth of science!

On the continent a different trend in philosophy emerged which was, on the face of it, much closer to science. Immanuel Kant was profoundly impressed by Newton's work on gravitation and was especially fascinated by his concepts of space and time. These formed the setting for his 'ideas of pure reason'—a realm far removed from the study of doubtful ordinary matter and familiar terrestrial happenings.

LOOKING OUT TOWARDS THE MILKY WAY

In the field of astronomy the early eighteenth century had developed some exciting new ideas about the fixed stars beyond the solar system. Christiaan Huygens seems to have made the first realistic attempt to estimate the distance of a particular star. He pointed his telescope at the Sun and then stopped down the light entering his instrument by passing it successively through two tiny holes until he judged it no brighter than the light of Sirius, the prominent 'Dog-star' of summer evenings. By this means he could calculate that, if Sirius had the same intrinsic brilliance as the sun it must be nearly 30,000 times as far away. It seemed a stupendous figure at the time, but now we calculate the distance of this close neighbour in space at over 700,000 times the distance of the sun. As this is already some 93 million miles the figures involved in measuring the distance to even the nearest of the stars becomes very unwieldy. For this reason a new unit of astronomical distance has been devised, the 'light-year' which, as its name implies, is the distance that a ray of light can travel in one year moving at its amazing speed of 186,000 miles per second. One light year is equivalent to nearly six million million miles so that the distance of Sirius from our earth can be more simply given as about *eight light years* in comparison with the Sun's distance of a mere *eight light minutes*.

Once the stellar Universe had been stretched so wide a new way of looking at the Milky Way became possible. This streak of diffuse light across the sky is a magnificent sight on a clear night and through a telescope appears spotted with thousands of faint stars. Perhaps the whole of its light is due to millions of distant stars, each too dim to see individually but together producing this path of pale light totally enclosing the earth? Perhaps, too, the Sun is no more than one of this huge swarm of stars? Kant, with his great admiration for Newton's work, saw this as the result of the force of gravitational attraction which he extended from the relatively close, familiar region of the planetary orbits, into the vast field of the stars themselves. Just as the planets revolve about the Sun in a series of ellipses all in the same plane, so the super-system of stars would form an immense flat elliptical shape moving round its own centre with such apparent slowness as to be undetectable when viewed from this distance. This circling movement had to be assumed to explain why the whole galaxy of stars had not long ago collapsed inward into one final gigantic heap—the centrifugal force of this motion would counter-balance the pull of gravity and so keep the whole system stable. If then the Earth were placed *within* such a disc-like distribution of stars it would be bound to appear to us like a streak across the sky.

Kant did have a little more evidence for his speculation since, a few years earlier, it had been shown that the 'nebulous stars'—like those in Orion and Andromeda—were both very faint and of measurable elliptical size. (All other stars appear as tiny pin-points of light even through the biggest modern telescopes.) These 'hazy' objects Kant took to be other Milky Ways far away in space but similar to our own galaxy. Later measurements were to prove these guesses partially correct but this alone was not enough for Kant. He had an ambition to show an immensity in time to match the splendour of his infinite space.

THE EVOLUTION OF STARS

The eighteenth century saw the beginning of time theories in geology, history, and biology. The sacred words of the Bible were no longer as binding on men's thoughts as they had been and the traces of evolution in rocks, civilisations, and animal species were stirring ideas about the age of things on Earth. Out in the enormous reaches of space Kant conceived a picture of cosmic evolution in which it was the active force of universal gravity that had first caused stars

FEBRUARY 1st.
10 P. M.

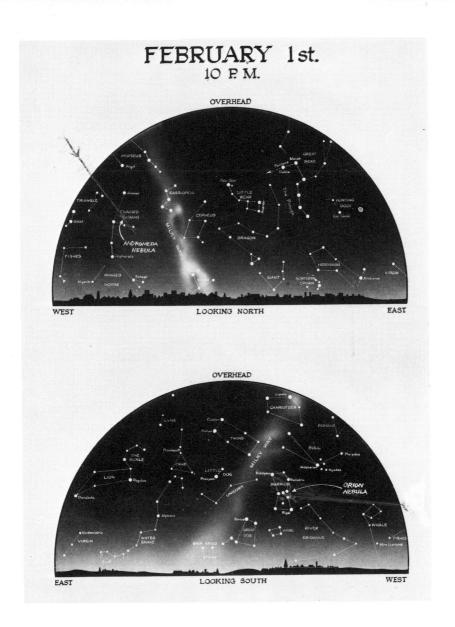

The stars arrive at the positions shown about half an hour later for each week before the above date

to condense out of primitive, chaotic matter and to form the great galaxies slowly throughout millions of centuries, only for them to fall gradually inwards again and to end their vast epoch of life in a final blaze of destruction. Yet this was no classical Armageddon, for Kant also held that, once the stars were exploded and their matter flung far out into space, the cycle of stellar evolution would begin again from the dust of this devastation :

> . . . matter which has to serve as material for worlds which are to be produced in the future, and of impulses for bringing it into motion, which begin with a weak stirring of those movements with which the immensity of these desert spaces are yet to be animated. (from *Universal Natural History and Theory of the Heavens.*)

For Kant there was no end to this process nor to the 'infinity in future succession of time'. His views on space and time were magnificient, inspiring, as he said himself, a full measure of 'silent wonder'. Space, it seemed was infinitely extended although it did, like Newton's space, have a fixed centre—the point where matter first formed stars. Time, too, was without end although it also had a beginning at the moment of divine creation.

Kant's cosmology was only loosely related to science and, in the hands of his followers, its divorce became more apparent as the eighteenth century continued to demonstrate its preference for abstract thought and 'enlightened reason' over the hard practical discipline of real science. So satisfied with this approach were some philosophers that they could write '. . . thanks to the Science of Astronomy we know the heavens greatly better than we do the Earth. . . . We shall have sooner arranged a system of fixed stars, and ascertained their motion than reduced the charges of the weather and the variations of the barometer to any fixed and determined rule.' But Nature was not to be so easily 'arranged'! In the year 1781 the great English astronomer William Herschel discovered the planet Uranus so bringing the total up to that philosophically satisfying and magical number—seven. While others continued to search the heavens for further planets the exponents of 'human reason' decided that such a significant number must denote completeness and poured scorn on those who wasted their time looking for more. The first year of the nineteenth century was marked by the publication of just such a sarcastic tract by Hegel and simultaneously the embarrassing discovery of the first of many minor planets—Ceres! The awkward liaison between science and philosophy which seems to have so hampered the growth of physics gradually faded away after this resounding public blunder and the new century which was to so transform man's ideas on light, force, and space itself, began its work.

8
The New Wave Theory of Light

The first subject to be studied afresh was light and the originator of the new approach was Dr Thomas Young who became Professor of Physics at the Royal Institution when he was only twenty-eight. His first topic was the puzzling colours on thin films which had forced even Newton to concede the existence of waves or vibrations alongside his light corpuscles. Young saw at once a possible reason for the production of these colours which does not seem to have been explored before. In films there are always two rays formed from each original ray that falls on the thin layer, one reflected from the top and one from the bottom. Now if light is 'undulatory'—to use his own word—it should be possible for two such waves to be 'out of step' with each other so that the 'crest' of one corresponded

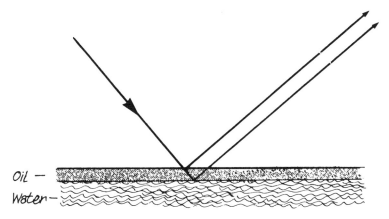

The 'splitting' of a ray of light by reflection at a film of oil on water

with the 'trough' of the other and thus cancelled out their effects completely. In such a case there should be darkness. But if two waves were 'in step' with each other they should add up to give an extra bright beam of light. As Newton had conclusively shown white light to be a mixture of pure colours the total destruction, in this way, of one colour would result in the appearance of its opposite or 'complementary' colour. White light robbed of red looks green: white without violet looks yellow, and so on. In this way Young

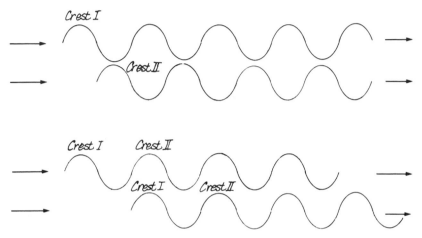

Waves out of step (phase) by half a wavelength, and waves 'in step' though moved along by one whole wavelength

managed to account for the colours on soap bubbles, films of oil or petrol on water and, ironically, even the so-called 'Newton's Rings' mentioned before by using the rival Wave Theory of Light. The basic phenomenon was called 'Interference' and Young set up many new experiments by which it could be demonstrated.

Young explained interference by analogy with sound waves producing 'beats' and water waves spreading out from two centres near each other. Circular ripples like these show the effect very clearly. Wherever the ripples from A and B are half a wavelength

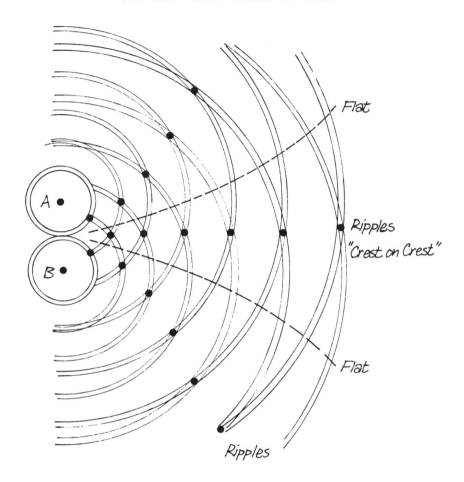

out of step the resulting interference leaves the water quite calm and undisturbed as though no waves at all were moving through it. A similar demonstration with light would be powerful evidence for Young's theory, but it was more difficult to achieve. Both beams of light must have the same origin in order to interfere with each other. In Young's most famous experiment he managed to make light spread out (diffract) from two tiny holes pierced very close together

92

in a dark screen. The resulting pattern showed the clear bright and dark fringes that he had expected.

> The combination of two portions of white or mixed light when viewed at a great distance, exhibits a few white and black stripes . . . although, upon closer inspection, the distance effects of an infinite number of stripes of different breadths appear to be compounded together, so as to produce a beautiful diversity of tints, passing by degrees into each other. (from Young's *Course of Lectures on Natural Philosophy and the Mechanical Arts*—1807.)

A practical result of this experiment was that it becomes possible to calculate the wavelengths of all the different colours by measuring the width of the stripes. Young found the red end of the spectrum to be composed of longer waves than the violet end, being about $1/36,000$ inch and $1/60,000$ inch respectively, with the other colours forming a continuous series in between. But the theoretical result of these experiments was more explosive—it threatened the very fabric of Newton's corpuscular theory of light. It is impossible to imagine how two separate 'very small Bodies' could possibly 'cancel' each other out by being out of phase so the whole theory of interference could not but be a straight challenge to this accepted view of light. But British science had revered the words of Isaac Newton for a full century and to many older and more conservative scientists the new theory seemed almost a blasphemy. One noble lord launched a thunderous attack on Young, accusing him of attributing 'presumed errors' to Newton—an inconceivable idea! Young wrote a reply maintaining, 'much as I venerate the name of Newton, I am not therefore obliged to believe that he was infallible'; but no one would even publish it. Blind prejudice had its day and for a while Newton's theory was propped up by orthodoxy but its time was now rapidly running out.

THE WAVES SPREAD ROUND EDGES

In France, where the Newtonian tradition was a little less strong than in England, Young was to find a formidable ally. Augustine Fresnel was an engineer but also a graduate of the illustrious new École Polytechnique and his imagination was caught by the study of light. He applied Huygens' elegant wave constructions to all the complicated series of coloured fringes formed within and without the edges of fine shadows. These beautiful effects can sometimes be observed when a bright street light is seen through the fabric of an umbrella and a host of tiny coloured spectra appear to stretch out from the light in two perpendicular directions, or when sunlight filters past one's eyelashes through half-closed eyes. The Academie des Sciences in Paris was so disturbed by Fresnel's flow of original papers on this topic that, in 1817, they inauguarated a public competition on the nature of diffraction. Probably they hoped for a rebuttal of Fresnel's treatment in which Young's hypothesis of interference was so successfully allied to Huygens' Wave theory of light; but they were to be disappointed. Apart from Fresnel there was only one other entrant and the former's work was outstanding. Every prediction from his theories could be demonstrated experimentally and in the end the Academie was forced to present him with the coveted prize. This moment (in 1819) virtually marked the end of the age of Newtonian Optics—the wave theory of light now seemed unassailable.

NOT COMPRESSION-WAVES BUT RIPPLES

At the same time Fresnel was studying the asymmetrical effects of crystals on a ray of light. This is the phenomenon which gives rise to the double images seen through Iceland Spar, for which Huygens had invented elliptical wavelets and Newton had given his corpuscles flattened sides. Fresnel discovered to his astonishment that the two emerging rays would not give interference patterns if recombined together. When Thomas Young was told of this he proposed a simple but revolutionary modification to his wave theory. He suggested that the light was not a *longitudinal* wave in which a compression travels through the aether, but a *transverse* wave in which a sideways ripple spreads through the medium.

Longitudinal (or compression) Wave (Huygens' type)

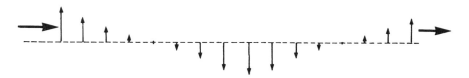

Transverse Wave (Young's type)

The relevance of this to doubly refracting crystals was that one transverse wave might well be resolved into two component rays in which the transverse vibrations were at right angles to each other if these could travel at different speeds through the crystals. Unlike liquids, glasses or normal metals, crystals do have differences in their properties at different orientations to their axis of symmetry. They are hard in some directions but split easily in others and, however they are formed, the angles between their faces never change; so it was not difficult to believe that the perpendicular vibrations of a transverse wave might meet with different resistances to motion at different orientations within the crystal. A longitudinal wave, on the other hand, could never be split into two in this way since it has only one direction of vibration. Young compared light waves to those that travel down a rope when it is flicked along the ground and he saw at once that two such 'polarised' waves in which the oscillations

95

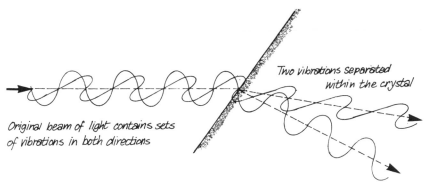

Two vibrations separated within the crystal

Original beam of light contains sets of vibrations in both directions

Two beams of light 'polarised' at right angles to each other by refraction in a crystal

were at right angles could not possibly interfere with each other. This made Fresnel's awkward experiment entirely understandable.

All subsequent work with polarised light, whether from crystals, or from reflection off glass and water, or from the blue light of a clear sky bore out Young's hypothesis completely. In these few years almost all the outstanding questions in optics seemed to have been cleared up and explained on the new Wave theory but, like the dust which is thrown up by energetic spring-cleaning, the problems tidied up from light settled obstinately in the more abstract field of the aether!

HOW DOES THE AETHER RIPPLE?

It is easy to see that transverse waves can only travel through a medium which has tension like a taut rope or an elastic solid. They cannot exist in either liquids or gases as both of these lack the essential property of springing back into shape. The embarrassing result, then, of Young's successful theory of light was that the age-old aether, through which the stars and planets moved without let or hindrance, had to be a *solid*— a kind of weightless jelly with such tension that it could transmit ripples at frequencies of more than a million cycles per second. It stretched belief to the limit! Only the heroically confident scientists of the nineteenth century could have taken up such a forbidding challenge with such dedicated seriousness.

The age of the amateur was over, the rift between philosophy and science was almost total, and the scientists of the Steam Age began to evolve their own style of work which was new and forceful. Perhaps it was because of the very machinery which contemporary progress and invention had brought into being that these men were so intent on producing 'models' for every new hypothesis. In biology, for example, theories of evolution had already been suggested in the eighteenth century but it took Charles Darwin in the nineteenth century to propose the mechanism of Natural Selection by which it could be seen to *work*. The growth and decay of civilisations had also been studied before, but it was Karl Marx who invented the model of exploitation by capital to give it a *method* of change. The physicists and mathematicians worked in a similar way using mechanical analogies for every effect in nature from the vibrations of an 'elastic foam' for describing the luminiferous aether, to stretched 'tubes of force' to account for magnetic attraction. Each model was analysed with mathematical virtuosity and used to make new and often far-reaching predictions. No wonder that the air was full of Victorian self-confidence!

9
Strains, Stresses, and Electric Waves

The most exciting field of study was certainly Electricity. At last the electric battery was available to produce steady currents and one of the first results of this was the discovery of an unsuspected connection between electricity and magnetism by the Danish physicist, H. C. Oersted. As soon as his results were published in England a young laboratory assistant at the Royal Institution— Michael Faraday—began the researches which were to be his life-work. At the same time the great French mathematician, A. M. Ampère, launched a new electrical theory based on the established model of charged fluids, but Faraday knew little mathematics and trusted no abstract theory; he was always to rely throughout his work on a carefully planned series of experiments. Through these he not only discovered a host of new electrical effects but he groped his way, by patient trial and error, towards a new conception of force and space that was more universal than any man before him had dared to dream of, let alone pursue by relentless experiment.

Ampère showed that two parallel wires carrying electric currents exert a force on each other. Was this more Newtonian 'action at a distance'? Oersted had shown that such a current deflects a magnetic compass needle and there was much public debate on how a flowing, spiralling or vibrating electric fluid could possibly produce such a magnetic effect. Faraday was intrigued by the direction and nature of this magnetic force. If an ordinary magnet is placed under a sheet of paper sprinkled with iron filings these arrange themselves in a series of well-known and characteristic lines, but the curves near a

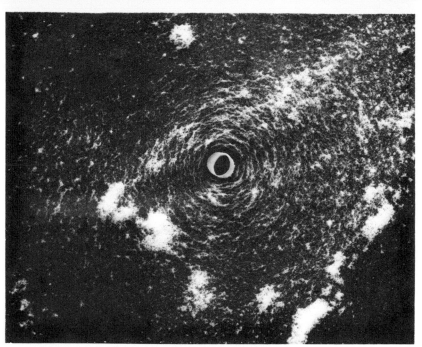

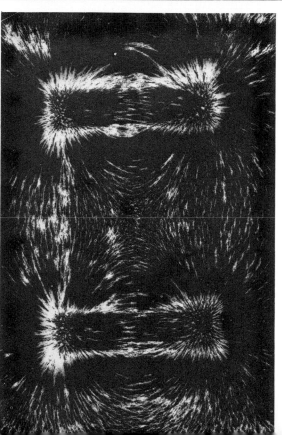

The negative print (left) was obtained by sprinkling iron filings around magnets as Faraday did himself. To the right a copper wire carrying a current was threaded through the central hole and iron filings showed fainter but quite definite circular lines of magnetic force

straight wire carrying a current bend round into perfect circles. There were no 'north poles' or 'south poles', just these curves of magnetic pull in the space around the wire. Once Faraday had thoroughly tested and understood the situation he designed a simple experiment in which a freely suspended wire carrying a current would be able to spin round and round a fixed magnet by the in-

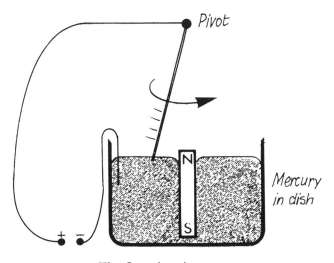

The first electric motor

teraction between these circles and the force of the magnet's north pole. The device was immediately successful and thus, in 1822, the prototype of the first electric motor was invented.

ELECTRIC CURRENTS FROM MAGNETISM

Throughout the next thirty years of experimental research Faraday gradually came to visualise these immaterial 'lines of strain' due to both magnetism and electricity. He guessed that the lines of force from one circuit, or from a magnet, would be able to induce

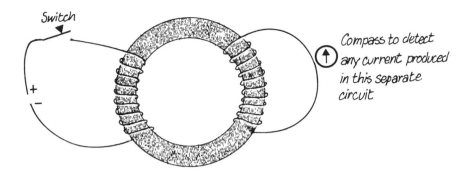

Faraday's ring. When a current was switched on in the left-hand circuit a pulse of current was induced in the right-hand circuit

electric currents to flow in a neighbouring circuit. He searched for this effect in 1825 but did not find it until more than six years later. The very first transformer and the very first dynamo were the practical results of these experiments, but in his own mind the rewards

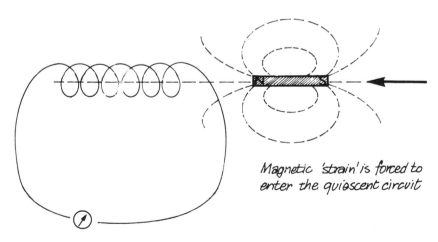

Current is induced by an approaching magnet whose 'lines of force' thread the circuit

were only a little more understanding and more clues and queries to follow up with yet more experiments.

He proved that it was the change in the magnetic 'lines of strain' linking the circuits that caused a current to flow. He looked for this strain too within insulating substances and discovered the powers of dielectrics which we use in modern capacitors for practical electronics. He also studied the movement of substances in solutions during electrolysis and found that it was not a case of remote attraction from the poles (terminal wires), but just a steady flow of charged particles along the familiar lines of force within the liquid. Then, because he wanted to demonstrate this internal strain in a

Electroplating by electric force within a liquid

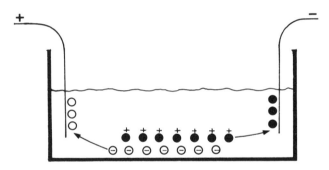

Charged ions moving as the 'strain' overcomes molecular forces in a solution

more striking way, he looked for an effect on *light* by either magnetic or electric strains and after many failures he found it. He sent a beam of polarised light through a glass rod in an intense magnetic field and showed, to his delight, that it was rotated by the 'diamagnetic' strain in the glass. Faraday reported his results under the title 'On the magnetisation of Light and the illumination of magnetic lines of force'.

I believe that, in the experiments I describe in the paper, light has been magnetically affected, ie that that which is magnetic in the forces of matter has been affected, and in turn has affected that which is truly magnetic in the force of light. (*Electrical Researches*, 1845.)

LINES OF STRAIN INSTEAD OF ACTION-AT-A-DISTANCE

When Faraday began his work in 1820 there had appeared to be only three main avenues of approach to the forces of Nature:

(*a*) Pure, unexplained 'action at a distance' as advocated by the followers of Newton and Coulomb.

(*b*) Vague 'imponderable' electric fluids, as Franklin had suggested, which filled all matter and space itself.

(*c*) The all-pervading aether which vibrated as beams of light passed through it.

But Faraday could never fully believe in these strange undetectable fluids and indeed one can see just how little appeal they must have had to so practical a man. Hesitantly at first and then with

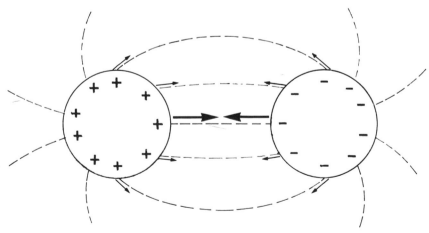

Force of attraction is the tug of lines of force

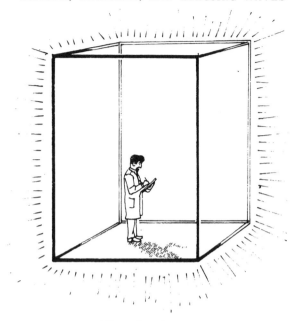

The 'electric cage'

increasing confidence he began to publish his own theory of 'lines of strain'—the closest description he could make to the actual forces that were experienced in the laboratory. The lines themselves were not a new discovery but Faraday's use of them was quite revolutionary because, for the first time in the history of science, it was suggested that the seat of the forces between one object and another was actually *in the space between them*. To Faraday this space—substance or vacuum—was full of tight tubes pulling out electric charges or magnetic poles on the objects to which they were attached at either end. It was a complete reversal of the usual line of thought where the charges or poles were seen as the active agents causing remote powers to act through neutral space. It must have seemed a very unlikely approach to his fellow scientists. In order to convince them, and himself, Faraday had to use all his inventive genius designing experiment after experiment to test and illustrate his new ideas.

For one public lecture Faraday built a huge cage of copper wire and tinfoil, 12 feet high, inside which he could stand and take measurements. The cube was charged with electricity until it sparked off every corner and yet Faraday detected no sign of charge on the inside. He hoped to prove that the enormous quantity of charge on the outside could not be due to concentrated 'electric fluid' but only to the lines of force from the walls of the room along which the crackling discharge took place. The spectacle was a great success but it did little to sway scientific opinion.

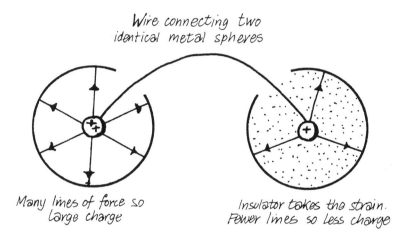

Wire connecting two
identical metal spheres

Many lines of force so
large charge

Insulator takes the strain.
Fewer lines so less charge

Faraday demonstrated quantitatively that the medium through which the lines of force passed affected their concentration. This again could *not* be explained on the classic fluid theory.

He tried to prove that electric charges were only pulled out of a substance by the tension of these imaginary lines of force which became so real to him that he could compare them to the filaments of a spider's web. He lowered a charged ball into a metal bucket and his instruments showed that exactly the same charge was induced in the bucket wherever the ball was placed. The number of

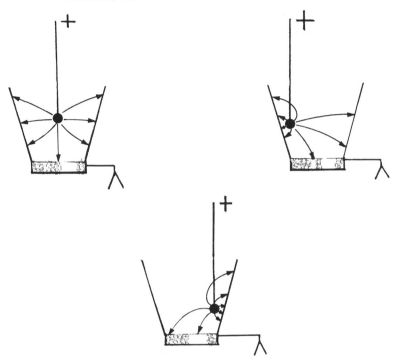

Charged ball produces the identical charge on the bucket in every case since all its lines of force terminate on the bucket

lines of force was the same so that they excited the same charge. It was quite the opposite effect to that expected by the proponents of the 'action-at-a-distance' school of thought.

UNITING LIGHT AND ELECTROMAGNETISM

Once Faraday had established that even light was influenced by his magnetic lines of force he became inspired to hope that the luminiferous aether itself might be replaced by these lines of tension along which the transverse waves of light would travel like ripples down a stretched string. Time was to prove the worth of this idea, too.

The view which I am so bold as to put forth considers, therefore, radiation as a high species of vibration in the lines of force which are known to connect particles and also masses of matter together. It endeavours to dismiss the ether, but not the vibrations. The kind of vibration which, I believe, can alone account for the wonderful, varied, and beautiful phenomena of polarisation. . . . (*Thoughts on Ray-vibrations*, 1846.)

Faraday hoped to include gravity, too, in his total scheme. He saw very clearly that, without lines of strain, the sudden emergence of a powerful force of attraction just when a second object was placed near the first was hard to accept. But if every object fills the space around it with *lines of gravitational tension* extending throughout the whole universe these would be ready to pull at any object as it approached. The location of just such a 'field of force' in the space around all massive objects was to be a fundamental feature of Einstein's Theory of Gravitation in the next century, but meanwhile Faraday, the tireless experimenter, searched for some single effect to show the hoped-for connection between electricity and gravity. By raising and lowering large insulated spheres of lead within the Shot Tower near Waterloo Bridge he hoped to detect electric charges induced by increasing gravitation of tension. Though these last experiments, performed when Faraday was in his late sixties, were not successful, they do illustrate the magnificent ambition of his life-long hope for a unity of all forces of Nature.

MAXWELL'S PREDICTION OF ELECTROMAGNETIC WAVES

Faraday's life was a success story in the best Victorian tradition—a poor boy makes good by deserving hard work!—and yet his conclusions proved, in their very simplicity, to be far ahead of his times. Already in 1855 a brilliant young Scottish physicist, James Clerk Maxwell, was beginning to clothe Faraday's ideas in a mathematical form but, in addition to all his elegant and difficult equations, Max-

Faraday giving the Christmas Lecture at the Royal Institution. The Prince Consort can be seen in the seat of honour accompanied by two of his young sons. Science was thought to be both fascinating and fashionable

well felt a need for the usual kind of mechanical analogy to describe the simple lines of tension through space that had satisfied Faraday. Forces and empty space had always proved uneasy bedfellows! At first Maxwell used a model of fluid flowing through 'tubes of force' and then, by 1861, he had constructed an enormously complex whirling-honeycomb structure for the very aether that Faraday had hoped to supersede, in order to analyse the electromagnetic effects that Faraday himself had discovered! Such an arrangement of the aether could be 'stressed' and distorted by electromagnetic forces but even Maxwell seems to have been aware how contrived this model must appear. He referred to it only as 'mechanically conceivable' and emphasised the 'provisional and temporary character of this hypothesis'. It was a device which filled the uncomfortable void with mechanical rotations like the great industrial machines of the age and it had, for Maxwell, a very definite purpose. Faraday's experiments had shown that moving charges (current) produce magnetic fields and that magnetic fluctuations can in turn produce electric currents. Did this mean that a kind of chain-reaction was possible in space? Could such a series of linked electric and magnetic disturbances ripple through the aether at a definite speed? This is the problem that Maxwell set himself: 'To find the rate of propagation of transverse vibrations through the elastic medium of which the cells are composed.' He first transformed Faraday's experimental results into a set of equations for three-dimensional variations of

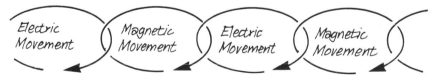

A 'chain' of electric and magnetic effects

electric currents and magnetic fields and from these he was able to obtain a recognisable wave-equation. He then substituted numbers from simple electric and magnetic measurements into his formula for the speed of such a wave—and it stood revealed as the well-known speed of light itself!

Maxwell let the numbers speak for him:

Predicted speed of hypothetical
Electro-magnetic waves   = 310,740,000,000 mm/sec
           or   193,088 miles/sec.

Most recent measurements of
the Velocity of Light   = 314,858,000,000 mm/sec
(at that time)   or   195,647 miles/sec.

The 'fit' was well within experimental error and Maxwell added his jubilant conclusion in italics:

> We can scarcely avoid the inference that light consists in the transverse undulations of the same medium which is the cause of electric and magnetic phenomena. (*On Physical Lines of Force*, 1861.)

Maxwell's masterly analysis of electromagnetic waves in the following years explained all the properties of light and colour in quite a new way. In 1873, when he published his famous *Treatise on Electricity and Magnetism*, he illustrated the directions of the rapidly alternating electric and magnetic fields producing the 'light' wave in a much simpler way. Yet it was hard to believe that white-hot luminous substances really did trigger off these electromagnetic disturbances at such high frequencies, and that our eyes could detect and 'see' them as familiar light. It would all have been much more

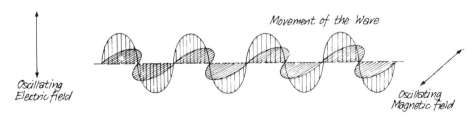

Movement of the Wave

Oscillating Electric field

Oscillating Magnetic field

convincing if the waves could be produced directly by electrical oscillations, and indeed the production of such 'radio waves' was now a predictable possibility.

THE WAVES ARE DISCOVERED

It was known at that time that the sparks from the discharge of a Leyden Jar, or other large two-layered metal object, were accompanied by a short burst of electrical oscillations, but it was years before anyone succeeded in proving that such surges of electricity could generate the waves that Maxwell had predicted. The man to whom the honour of solving this problem is due was Heinrich Hertz, Professor of Physics at Karlsruhe in Germany.

Hertz was a follower and great admirer of Maxwell and had long cherished the idea of a direct, experimental identification of his

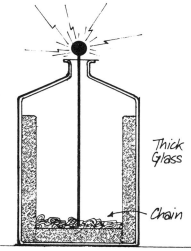

Leyden Jar. A glass container coated with metal foil and fitted with a central metal knob which was much used for storing electric charges like the modern capacitors inside a television set

electromagnetic waves. Sometime in early 1886 Hertz found a large pair of similar insulated spirals in his laboratory and decided to use them in his next tests. He connected one of these spirals across the knobs of a discharging Leyden Jar and saw faint sparks flashing between the knobs of the second spiral which was some distance away. Was this a simple electrostatic effect produced by forces acting at a distance, or was it the result of a travelling wave? Fortunately there was a straight piece of metal near by which was also being influenced by the rapid surges of electricity in the first spiral. Hertz found that

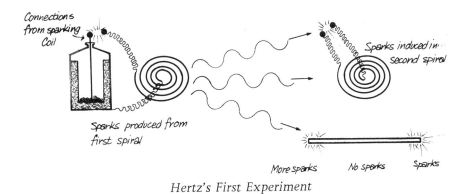

Hertz's First Experiment

though sparks could be easily obtained from either end of this rod, its mid-point remained obstinately neutral. It was like the swinging ends of a see-saw rocking on its stationary pivot and this alone was all the evidence Hertz needed to be sure that it was indeed electrical oscillations which were being transmitted. He then set about improving his apparatus so that continuous waves of a definite size could be generated and received over greater distances.

This time he eliminated the massive Leyden Jar so that the two circuits, transmitter and receiver, became almost identical, 'tuned' to the same frequency of electrical oscillation for resonance to occur between them. It was like a singer and a wine-glass—if the note emitted by one exactly matched the natural frequency of the other the response of vibration would be very much greater (to the point of shattering the glass in the case of sound!). This method worked admirably. Hertz experimented with a large number of different oscillators in place of the original spirals and obtained different frequencies of electrical vibrations and different corresponding wavelengths. Most of his experiments were with waves about 50cm long at a frequency of about 600 million swings per second. His carefully tuned receiver resonated giving tiny visible

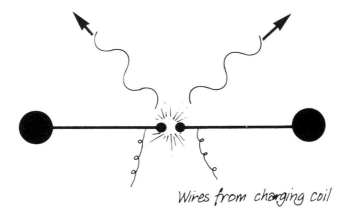

Wires from charging coil

The First Hertz Oscillator

sparks no more than a few millimetres long but, with this simple apparatus, he was able to prove that these electromagnetic waves did everything that light waves could do. They were reflected by metal mirrors, refracted by great prisms of pitch, polarised by metal gratings, and gave all the interference and diffraction patterns that were known for light. Above all, Hertz was able to prove that they travelled with a speed identical to that of light waves as Maxwell had predicted more than twenty years earlier. It was true that they did not affect the human eye and that their wavelength was more than a million times greater, but the family resemblance was undeniable. In practical terms Maxwell's obscure equations had been brought to life creating those 'wonderful wire-less waves' which were to give us radio, radar, television, and radio-astronomy in the present century.

10

The Problem of the Aether

Hertz was much more than an ingenious experimenter, his ambition was no less than the perennial inspiration of all scientists—the unification of all the phenomena of Nature. When his practical work was complete he delivered a famous lecture on 'Light and Electricity' in which he used his results as conclusive evidence of the identity of light and electromagnetic waves.

> All these experiments in themselves are very simple, but they lead to conclusions of the highest importance. They are fatal to any and every theory which assumes that electric force acts across space independent of time. They mark a brilliant victory for Maxwell's theory. No longer does this connect together natural phenomena far removed from each other. Even those who used to feel that this conception as to the nature of light had but a faint air of probability now find difficulty in resisting it. In this sense we have reached our goal. (Hertz, 1889.)

But science is doomed never to reach any final goal and a man with the insight of Hertz could see clearly all the problems thrown up by the very success of Maxwell's theory which still called urgently for attention.

> We are at once confronted with the question of direct actions-at-a-distance. Are there any such? Of the many in which we once believed there now remains but one—gravitation. Is this too a deception. . . ? Directly connected . . . is the great problem of the nature and properties of the aether which fills space, of its structure, of its rest or motion, of its finite or infinite extent. More

and more we feel that this is the all-important problem, and that the solution of it will not only reveal to us the nature of what used to be called imponderables, but also the nature of matter itself and of its most essential properties—weight and inertia. (*op cit*)

What an astonishing prophecy! By intuition based on deep comprehension Hertz had accurately forecast one of the greatest revolutions that physics has ever known.

DOES THE AETHER MOVE?

The nineteenth century had struggled desperately to understand the aether ever since Young and Fresnel had established the wave nature of light. A wave, they thought, had to be a *disturbance* of *some medium* and several generations of mathematicians had

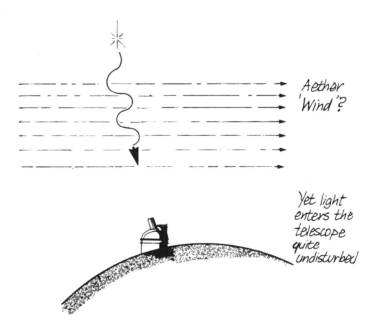

Aether 'Wind'?

Yet light enters the telescope quite undisturbed

wrestled with the problem, inventing model after model of this invisible, intangible stuff which was supposed to permeate both space and matter. Fresnel had used astronomical observations to show that light from a distant star was unaffected by movement of the intermediate aether. Young wrote 'that the luminiferous aether pervades the substance of all material bodies with little or no resistance, as freely perhaps as the wind passes through a grove of trees'. Yet, because transparent substances like glass slow down the speed of light, the concentration of aether within them had to be greater than in space. This theory earned some measure of support in spite of the absurdity of having to imagine different densities of aether within the same substance (or an infinity of different matters) to correspond with all the different colours of the spectrum! Maxwell, as we have seen, took the mechanics of the aether very seriously and imagined it being carried about within moving matter although later experiments seemed to disprove this hypothesis too. Towards the end of the century a last, almost despairing, effort was made to fit this stubborn material into the scheme of science.

By now the situation seemed to call for Alice's Humpty Dumpty who could believe three impossible things each morning before breakfast! The new theory, due to the Dutch physicist, H. A. Lorentz, beat a tactical retreat before the challenge to imagine anything at all about the aether. Lorentz hoped only to produce a mathematical scheme that would agree with all the experimental results in electromagnetism and yet assert no more than that all matter could move quite freely through this imperceptible stuff. He re-examined Maxwell's 'aetherial equations' for wave-motion and found that they still held good and so he trusted that his modest claim for the mere existence of the aether could not be experimentally disproved.

> Whatever it [the aether] may be one would do well, as I see it, not to be guided in such an important question by considerations of the degree of probability or simplicity of one or other

hypothesis but to rely on experiment for learning about the state of rest or motion in which one finds the aether. (Lorentz— quoted by Michelson and Morley, 1887. *Phil. Mag.*)

The challenge to experiment was taken up and the results proved both disastrous to Lorentz's theory and almost incomprehensible in themselves. A. H. Michelson and E. W. Morley, the architects of these researches, used two identical beams of light travelling along perpendicular tracks and observed the interference fringes obtained by combining them together again. They then carefully swung the

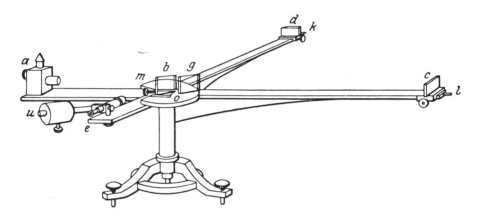

Michelson and Morley experiment—a contemporary drawing of the first apparatus to be used

whole apparatus through a right angle and once more measured the fringes obtained. Now if the aether which carries the light waves is drifting past the earth it is clear that it would speed up the passage of the light waves on one of the paths more than on the other and a noticeable shift in the interference pattern would result. They produced a piece of apparatus so delicate and sensitive that it had to be mounted on a stone slab floating in a pool of mercury to avoid the incessant vibrations of Berlin city life even at two o'clock in the morning. In spite of this no change whatever in the pattern could be detected. The unbelievable conclusion could now be drawn that the earth is at rest in the aether although we know we are moving continuously in orbit round the sun at a speed of 75,000 miles per hour! The 'luminiferous aether' seemed to elude every experimental probe. After nearly a century of continuous work on the subject absolutely *nothing* positive had been found out about it. It had no weight, no resistance, and no movement. The question would have to be faced— was the aether real?

DOES THE AETHER REALLY ALTER LENGTHS AND TIMES?

Lorentz and many other physicists of his time were unwilling to tackle such a fundamental, almost philosophical, question. But as disciplined scientists they could not ignore the facts of experiment so now it was essential to write into Maxwell's wave equations the results of the Michelson-Morley experiment that the speed of light was always the same regardless of the movement of the observer through the unobservable aether. It was a curious and unfamiliar task since the elusive aether's last remaining function was to carry these waves in the same way as still air transmits sound waves; but movement of air clearly does affect the speed of sound as anyone who has had to shout on a windy day will know. The contrary behaviour of the aether made the whole mechanism very hard to believe but then Lorentz, by his own admission, was now looking for neither 'probability' nor 'simplicity' and indeed the results he obtained were far from being either probable or simple.

First, it was clear that, however fast an observer pursued after a beam of light, it would always recede from him at exactly the same speed. That in itself was as mad as Alice's race with the queen in the Looking-Glass world; but when Lorentz examined the further implications of this idea yet more oddities turned up. We can forecast some of these results without recourse to complicated mathematics. Imagine an observer A, for Albert, who sets out to calculate the speed of light by measuring the *wavelength*, using a *ruler* and counts number of such waves passing him per second, the *frequency* by means of a clock. He will then multiply these two numbers together since Wavelength × Frequency = Speed of the Wave.

Now Albert firmly believes that he is at rest in the aether and if in doubt can always perform the Michelson-Morley experiment to prove it. He then sends out his friend B, for Bertie, to measure the

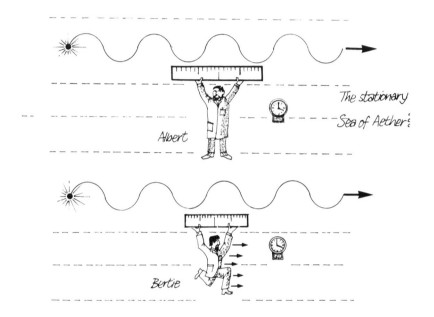

speed of light by the same method in a fast-moving laboratory. To Albert it seems clear that this tremendous speed through the aether 'squeezes up' all the light and shortens its wavelength. Also he knows that Bertie should get a different, lower, frequency for the waves since he is moving after them so fast that fewer crests will pass him in a second. However, in Lorentz's new universe Bertie will find everything unchanged so that his speed through the aether remains undetectable. Albert, watching from his resting position, still holds the firm conviction that shorter waves and slower frequency did exist in Bertie's flying laboratory. He concludes that Bertie's instruments must have gone wrong! His ruler must have contracted and his clock must have run slow by just such an amount as to make the effects that Albert *knew* had taken place. It was a strange new universe.

These were just the results that Lorentz obtained by his mathematical analysis of Maxwell's equations. The speed of light in the aether could only remain constant if motion through the aether produced

(a) contractions in the length of every object, and
(b) the slowing down of every timing device used.

For the first time in the history of science a 'local time-scale' had to be specified to distinguish it from all others and it was not a result that Lorentz relished! The mathematical transformations that he established have brought him lasting fame but years later he still spoke nostalgically of 'the concept [which the present author would dislike to abandon] that space and time are completely distinct and that one "true time" exists' (1910).

So it came about that, after a century of virtual divorce between science and philosophy, the basic concepts of measurement and reality had to be re-examined so that the bewildered world of science could unravel the bizarre results of its own progress.

(1) What does one do about a substance that is totally unobservable?

(2) Are measurements of length and speed in space only relative to each other?

(3) Is time absolute, or is it too only relative to the motion of an observer?

It is well known that the man who answered these questions and used them to formulate his new Theory of Relativity was Albert Einstein.

II
Relativity – Speed and Time

Einstein wrote that he had been much influenced by the philosophy of Ernst Mach, a nineteenth-century mathematician, and David Hume, an eighteenth-century sceptical philosopher. Nevertheless he believed that the problems of physics called for a physicist and *not* a philosopher to solve them. He wrote, 'Today when experience forces us to seek for a newer and more solid base, the physicist can no longer simply abandon the critical examination of theoretical bases to the philosopher, for he knows and certainly senses better where the shoe pinches.' His reappraisal of the abstract ideas of space and time was empirical and constructive so that it led to a powerful new theory which could be tested quantitatively against experiment and observation.

Hume had been no scientist but he had held mathematics and experiment in great respect. On the other hand he expressed his views on vague speculation with typical eighteenth-century vigour—

> if we take in hand any volume: of divinity or school metaphysics, for instance; let us ask 'Does it contain any abstract reasoning containing quantity or number?' No. 'Does it contain any experimental reasoning concerning matter of fact and existence?' No. Commit it then to the flames; for it can contain nothing but sophistry and illusion. (from *Inquiry concerning Human Understanding.* 1751.)

Einstein may well have recognised the solution to the problem of the aether in this thunderous denunciation from the past. By Hume's test all the multitude of words written about the unobservable aether could be thrown on the bonfire, and the liberation of physics from the continual presence of this embarrassing fluid allowed Einstein a new freedom of speculation. In his theory each moving frame of reference was exactly equivalent to every other frame of reference in space, since no special 'resting aether' needed to be considered. In the opening paragraphs of his first paper on Relativity, Einstein dismissed the aether in the following words:

> The introduction of a 'luminiferous aether' will prove to be superfluous inasmuch as the view here to be developed will not require an 'absolutely stationary space' provided with special properties. . . . (*Annalen der Physik*, 1905.)

At this time Einstein was an obscure employee in a Swiss patents office and it was some years before his theory won the recognition it deserved. In spite of his reputation for incomprehensibility, the ideas on which he based his new scheme shine out with lucid **simplicity. In fact, he advanced only two principles, both of which** were suggested by experience:

(1) *The Principle of Relativity*
'The same laws of electromagnetics and optics will be valid for all frames of reference for which the equations of mechanics hold good.'
(2) (A complementary extension of the first principle)
'Light is always propagated in empty space with a definite velocity "c" which is independent of the state of motion of the emitting body.'

RELATIVE SPEED

The second of these is simply the same old Michelson-Morley experimental result that now, after twenty years, was becoming understood at last: waves in empty space can only travel at one speed when there is no 'aether wind' to blow them fast or slow. The main, first principle is an attempt to include Electromagnetism into a kind of relativity in which we have always instinctively believed. Sitting in a railway carriage and watching another train pass the window often gives the uneasy feeling that it might well be our own train which is moving backwards. Only by checking with the position of nearby fields and buildings or feeling for the jolting of the train can we be sure which movement is 'really' happening. Were we in smooth rockets passing in empty space no such criteria for 'reality' would exist. Inside each rocket exactly the same laws of mechanics would hold good, objects would fall or float, bounce or balance in the same way and provide no possible means of deciding which rocket was moving. Indeed if the question has no observable meaning why ask it? Then the search for *absolute* movement, as opposed to relative movement, becomes as illusory and futile as the search for the discarded aether.

Using these two principles alone Einstein could arrive at the famous 'Lorentz Transformations' without reference to either Maxwell's equations or the aether which had inspired them. All lengths would appear shorter to a moving observer—not because they contract when they move through the aether as Lorentz believed—but because light (or radar) would have to be used at some point in their measurement. Similarly clocks would run fast or slow depending on the speed of the observer because a light beam from one would be needed to synchronise the other. This was a consequence that, unlike Lorentz, he fully accepted.

RELATIVE TIME

Einstein held that universal standard time existed no more than absolute speed or absolute length. This was a purely philosophical point that had been made in criticism of Newton's classical mechanics twenty years earlier by Ernst Mach. He had written that however intervals of time are measured, by the swinging of a

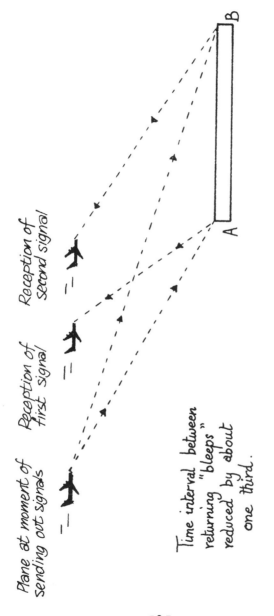

Measuring a rod by radar during flight. In this exaggerated drawing the 'plane is travelling towards the returning signal and will receive it sooner than if it had been at rest and so will judge the rod to be shortened

pendulum, by the rotation of the earth, or by the beat of a human pulse, none of these methods can give any inkling of what Newton had grandly called 'Absolute, true, mathematical time, [which] of itself and by its own nature flows uniformly on without regard to anything external'. Nothing external whatever? . . . of what use, then, is any clock?

> . . . time is an abstraction, at which we arrive by means of the changes in things; made because we are not restricted to any one *definite* measure, all being interconnected. . . . But the question whether a motion is in *itself* uniform, is senseless. With just as little justice, also may we speak of an 'absolute time'—*of a time independent* of change. This absolute time can be measured by comparison with no motion; it has therefore neither a practical nor a scientific value; and no one is justified in saying that he knows aught about it. It is an idle metaphysical conception. (Mach, *Science of Mechanics*, 1883.)

Einstein could go further than this by using his Principle of Relativity. He showed that the measurement of time is so intimately related to the speed of the point of reference that even events which might appear to be exactly simultaneous to one observer may seem to be happening at different times to another who is moving. Einstein's own illustration of this extraordinary result cannot be bettered for clarity.

The 'Relativity Express'

We suppose a very long train travelling along the rails with the constant velocity V and in the direction indicated in [Fig below]. People travelling in this train will with advantage use the train as a rigid reference body (coordinate system); they regard all events in reference to the train. Then every event which takes place along the line also takes place at a particular point of the train. Also the definition of simultaneity can be given relative to the train in exactly the same way as with respect to the embankment. As a natural consequence, however, the following question arises:

Are two events (eg the two strokes of lightning A and B) which are simultaneous *with reference to the railway embankment* also simultaneous *relatively to the train*? We shall show directly that the answer must be in the negative.

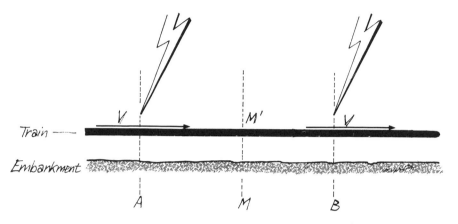

When we say that the lightning strokes A and B are simultaneous with respect to the embankment, we mean: the rays of light emitted at the places A and B, where the lightning occurs, meet each other at the mid-point M of the length A→B of the embankment. But the events A and B also correspond to positions A and B on the train. Let M^1 be the mid-point of the distance A→B on the travelling train. Just when the flashes of lightning occur (as judged from the embankment) this point M^1 naturally coincides with the point M, but it moves towards the right in the diagram with the velocity V of the train. If an observer sitting in the position. M^1 in the train did not possess this velocity, then he would remain permanently at M, and the light rays emitted by the flashes of lightning A and B would reach him simultaneously, ie they would meet just where he is situated. Now in reality (considered with reference to the railway embankment) he is hastening towards the beam of light coming from B, whilst he is riding on ahead of the beam of light coming from A. Hence the observer will see the beam of light emitted from B earlier than he will see that emitted from A. Observers who take the railway train as their reference-body must therefore come to the conclusion that the lightning flash B took place earlier than the lightning flash A. We thus arrive at the important result:

Events which are simultaneous with reference to the embankment are not simultaneous with respect to the train, and vice versa (relativity of simultaneity). Every reference body (coordinate system) has its own particular time; unless we are told the reference body to which the statement of time refers, there is no meaning in a statement of the time of an event. (Einstein—*Relativity: The Special and General Theory*, 1920.)

Such illuminating 'thought experiments' often played an important part in Einstein's approach to scientific problems. Perhaps it is characteristic of a man so unaffectedly modest, that he would publish such an unsophisticated argument along with his 'heavier' more impressive mathematical treatment of the subject. Of course he did derive a precise equation for determining the corresponding lengths of an interval of time as measured by two observers moving past each other in space which is, in its simplest form, as follows:

$$\text{Time interval for A} = \frac{\text{Time Interval for B}}{\left[\sqrt{1 - \frac{v^2}{(300,000,000)^2}}\right]}$$

The figure 300,000,000 is the speed of light in metres per second and v is the relative velocity of the two observers in the same units. It is true that none of us have ever seen real clocks behave in such a bewildering manner but if we examine the numerical relation more closely the reason for this becomes apparent. If A were installed in one of our fastest aeroplanes he would still be travelling at a mere 600m/sec. Substitution of this number into one equation shows that this will make his time-piece run slower than ours but the difference will be less than one second in a whole century because his speed is so slow when compared with the swiftness of light. Only if we were propelled at a truly fantastic rate, nearly a hundred thousand times as fast as this aeroplane, would the time dilation become perceptible. Unattainable as such a speed must seem it is *not* unobservable in practice.

EXPERIMENTAL EVIDENCE

Although man himself cannot yet travel at these incredible speeds there are minute sub-atomic particles which are produced naturally with velocities closely approaching that of light itself. So-called 'cosmic rays'*—which are probably hydrogen atoms travelling with enormous energy—collide with nuclei high up in the topmost layer of our atmosphere and give rise to very swift transient particles, called 'mesons', which can be detected by the vapour trails that they form in cloud chambers at ground level. Now when we produce such particles artificially by nuclear bombardment they spontaneously break up within two millionths of a second. This is their allotted life-span. The question is how they can possibly survive long enough ever to reach the surface of the earth. In spite of their great speeds one can calculate that it must have taken them at least twenty millionths of a second to travel the hundred-odd miles down to ground level from the top of the atmosphere.

Reference to Einstein's equation on time dilation solves the riddle. The average speed of these mesons is about 285,000,000 m/sec or 95 per cent of the speed of light and this would expand their time-scale, relative to us, by a factor of ten. The length of their own life-span, like all their 'clocks', will dilate by this amount and so allow them enough time to reach the earth, without disintegrating! Life-time of a meson at rest relative to us

$$= \frac{2}{1,000,000} \quad \text{second}$$

* An explanation of the terms used in the following paragraphs may be found in the author's *The Structure of Matter*, a companion to the present volume.

Time taken to traverse the atmosphere

$$= \frac{20}{1,000,000} \quad \text{second}$$

These experiments, which were not made until 1941, constituted a triumphant verification of Einstein's theory of time—if indeed one were needed by that date.

The reader may now enjoy translating the experience of this ephemeral speck of matter into human terms in order to imagine the strange possibilities of rapid space-flight! However, it must be

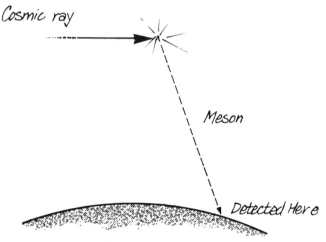

How can the meson reach the ground?

stressed that to the meson time will still seem to tick by its usual rate, though as it hurtles towards the earth, the depth of our atmosphere will appear so much contracted that the meson would naturally 'expect' to be able to traverse it without difficulty during its brief existence. By the operation of the Principle of Relativity no one reference system can ever seem strange to its own inhabitants—relativistic peculiarities, in common with 'motes in the eye' can only be detected in others! Every observer can believe himself (like the pre-Copernicans) to be at perfect rest in the only 'normal' world amid a plethora of deluded, speeding observers measuring 'wrong' lengths and 'wrong' times!

THE MARRIAGE OF SPACE AND TIME

Before the days of Relativity it was considered possible to identify any point in space by reference to three lengths, the so-called

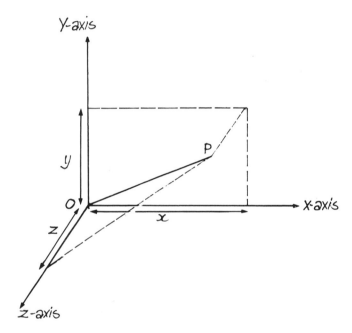

Fixing a point P in space by measuring the 3 coordinates, x, y and z

'space coordinates' which correspond to the three dimensions of height, width, and breadth. But now we know that such measurements would not 'fix' the point for any other observer who was moving past since such lengths might contract for him.

Similarly the time of an event used to be fixed by reference to an independent time-scale that was thought to be common to all observers. That, too, it seems, can no longer be done; time intervals, like lengths 'come out different' for different observers. In other words neither space nor time is absolute. Is there any quantity or collection of quantities that can ever again pin-point one event in our Universe? The answer given at that time was *yes*. Because the speed of all electromagnetic waves—radio, light, heat, or X-rays—is unchangeable and because speed involves the measurement of both length and time it follows that a combination of space-and-time can be used to identify any event as precisely as Newton's more familiar absolute space had done before. Now, however, not only are three coordinates of length required but a time measurement must also be included. It is in this sense that Einstein's universe is said to be four-dimensional. Such a remark, when shorn of its mystery and glamour, is seen to be no more than a recipe for describing 'a happening' in the universe to the satisfaction of all possible moving observers who can easily measure their relative velocity and 'transform' their own measurements according to the Lorentz-Einstein equations. This is the only universal picture of *valid interconnected experiences* that can now hold good.

RELATIVE MASS

To return to the spade-work of Relativity one can easily foresee that many of the other measurements made will vary with the speed of the observer. Accelerations and velocities, for example, depend on both the distance travelled and the time taken so that they too will 'transform' in new ways. Indeed Einstein set himself a large programme of work when he maintained that *all* the laws of mechanics, optics, and electromagnetism must be unchanged in every frame of reference. The laws of mechanics had been laid down by Isaac Newton 240 years earlier and these were the laws that Einstein had now to reformulate in his new scheme. As he examined the problem of colliding balls he realised that even the mass or 'inertia' of an object must be affected by its speed relative to the frame of reference.

He reasoned that when an object gains energy, whether it be by impact or from radiation, two general effects can follow :

(1) Its mass may increase
> and
(2) It increases its kinetic energy (speed)

No longer could chemists hold that the total mass of matter is always the same; no longer could physicists hold that the total

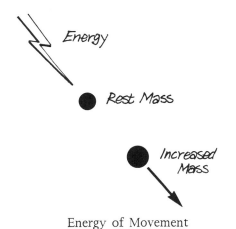

Energy of Movement

energy of a system is never diminished. Einstein's equations, backed by experimental evidence, showed that *some energy does disappear* only to produce *a corresponding increase in mass*, as though these two different entities were somehow interchangeable. It follows that the fastest particles will naturally 'weigh' more as a direct result of having so much energy of movement (relative to us). The old Conservation Laws of Mass and Energy needed to be welded together into one new law which would express their equivalence and totality.

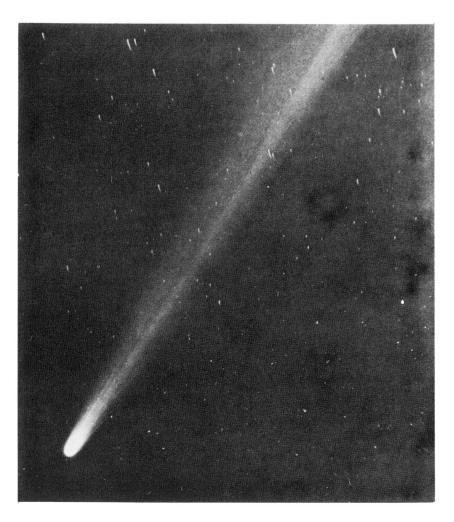

Photograph of Halley's Comet. Its long trailing tail is so thin that stars can clearly be seen through it and the impact of the Sun's energy is forcing it to stream out behind the comet's more massive head

The sum total of all the Mass-Plus-Energy in the Universe is un-changeable.

One similarity between 'solid' matter and 'pure' energy was already known. Intense light is capable of exerting a real mechanical pressure on objects as though it were a hail of tiny bullets. The tails of great comets, which show this effect, are made up of a stream of dust so fine and diffuse that the Earth has passed right through such a tail during this century without catastrophe or comment. The pressure of strong sunlight on this tenuous substance is sufficient to 'blow' it behind the comet away from the sun whichever way the comet may be moving. Tycho Brahe and Kepler had made this observation three hundred years earlier and Maxwell had offered some explanation on his Electromagnetic Theory but now it could be seen as a simple illustration of 'incorporeal' energy impinging on particles of matter and making them recoil just like a collision between two hard balls.

In one vital respect Einstein's theory went far beyond contemporary experiment. Even the mass that an object possesses *when at rest* relative to the observer must be equivalent to energy so it should be possible, under the right conditions, to annihilate some matter and produce energy in its place. Then :

$$[\text{Energy Produced}] = [\text{Mass eliminated}] \times [\text{Speed of Light}]^2$$

Since the speed of light is such an enormous figure this equation predicted a vast liberation of energy for even a minute loss of mass—about thirty million kilowatt-hours of power per gram of matter. Not until 1932 had nuclear physics progressed far enough to produce experimental corroboration of this remarkable prophecy. But now that the world has witnessed the shattering effects of atomic bombs, in which small quantities of matter are converted into awe-inspiring amounts of destructive energy, man can no longer doubt the validity of this sinister equation !

Such were the main results of Einstein's Theory of Special Relativity. It was a magnificent structure involving a whole new view of space and time, of matter and energy and, within a few years, it won an enormous measure of delighted acceptance. Of course it had not sprung suddenly, like Venus, fully grown from a sea of ignorance; it was, in one sense, the culmination of the work of Maxwell, Lorentz, Poincaré and others. Even the title 'Relativity' was not coined by Einstein, and yet there was a logical completeness in his argument which was tremendously new and convincing. To many contemporary physicists who had been struggling with the problems of electromagnetic waves in the aether, of time, length, and speed, it was almost a revelation. A Polish professor, on reading Einstein's first paper, is said to have announced to his students—'A new Copernicus is born !'

The man who brought about this twentieth-century revolution is no enigma for he is the subject of a large number of biographies and was himself a fairly prolific writer. It all makes fascinating reading for he lived through a period of wars, persecution, and social upheaval and, though he despised politics, he reacted to all the currents of thought of the day with characteristic energy of mind and simplicity of spirit. He believed that a sense of individual isolation was a characteristic of the times but that 'Man can find meaning in life, short and perilous as it is, only by devoting himself to society.' This statement from Einstein is particularly poignant since by his own account he suffered from just such a sense of isolation from early childhood, but he used this loneliness to look at the world with new eyes to the lasting benefit of science.

Einstein's formal education had not been a very happy process. At the beginning he had been the only Jew in a Catholic school and the only pacifist at a time of militaristic mania. It was from his home that he first learnt the wonder of science when, at the age of five, his father showed him a simple compass and later when an uncle bought him Euclid's *Geometry* and lent him popular books on science. This feeling of wonder, coupled with a delight in the logical precision of thought as displayed in mathematics, formed the living basis of his attachment to science as they probably have done for scientists in every century. It is fascinating to see how Einstein himself linked this personal reaction to the mystery of Nature with the power that has engendered religious emotion in man. He was a non-observing Jew in an age which saw the rise of Scientific Humanism and the decline of all organised forms of religion, yet he wrote:

> A knowledge of the existence of something we cannot penetrate, of the manifestations of the profoundest reason and the most radiant beauty, which are only accessible to our reason in their most elementary forms—it is this knowledge and this emotion that constitutes the truly religious attitude; in this sense, and in this alone, I am a deeply religious man. (from *The World as I see it*, Einstein, 1935.)

He says himself that he had been struggling with the concept of speed since the age of sixteen when he first faced the imagined paradox of chasing after, and even catching up with, a beam of light. He could see in his mind's eye, that, to such an observer, the light wave would seem like a stationary electromagnetic oscillation, which is not feasible. As he learnt more about the methods of physical investigation, he noticed particularly how powerful a principle which maintains the impossibility of a certain situation has sometimes proved. The whole science of thermodynamics, for example, was based on the statement—'A perpetual motion machine is impossible'. He had a hunch, he said, that some such negative principle would lead to a solution of the current problems of electrodynamics. In Faraday's work he could see the effects of electromagnetic induction as a result of *relative* motion between a magnetic field and a circuit. During ten years these different strands of

Head of Einstein from a sculpture by Epstein

thought worked out a pattern in his mind until, by 1905, he was able to recognise the negative principle he was looking for as— 'Absolute motion is impossible'. He combined this with a careful examination of how one actually measures time and length until he was in a position to guide the steps of the emerging Relativity along a sure, clear path. His success was such that 'Special Relativity' is now incorporated into so many branches of physics that it would be hard to imagine any experimental result that could cast serious doubt on its validity.

SCIENTIFIC OBSERVATION

However paradoxical some of its results may seem, Special Relativity bears the unmistakable mark of our own age. Absolute time and the aether were discarded because they were shown to be *unobservable*. Variations in lengths, times, and masses were predicted by examining the way in which they would be *measured*. Not by speculation, nor by experiment, but by an analysis of the significance of measurable concepts the new physics came into being. Ernst Mach and David Hume, who so influenced the young Einstein, also inspired a powerful movement in philosophy at the turn of the century which was called 'Logical Positivism'. This set out to reject, as Einstein did, all *a priori* ideas that smacked of metaphysics and insisted that the meaning of any statement was no more than the method by which it could be verified experimentally. At first this ruthless approach successfully cleared away a lot of meaningless cobwebs but finally it became sterile and destructive as it threatened the essential free play of human imagination. Einstein himself grew to regret the encouragement he had once given to this way of thought. Modern Philosophy is still much involved in 'epistemology' —the study of how knowledge is acquired—and within science this has become an examination of 'how observers observe'. Sir Arthur Eddington, who was an early advocate of Relativity, maintained that he could forecast the numerical results of some experiments by merely observing the method by which the research was being carried out. Ours is indeed a self-conscious age and psycho-analysis is only one aspect of the inner searching that has become such a characteristic of our method of inquiry.

12
Acceleration, Gravity, and Light

At this point Einstein was still only 26 and his most imaginative and universal creation still lay ahead. So far his Theory of Relativity was called 'Special' or 'Restricted' because it held good only in those systems between which there was steady, uniform speed. But once a system speeds up (accelerates) it is very easy to distinguish from all other frames of reference. In an accelerating room, for example, plates would spontaneously fall off the shelves, the lamp would hang to one side and the astonished occupants would be thrown against one wall and pinned down against it! In any non-accelerating room, just one such effect would appear as the wildest feat

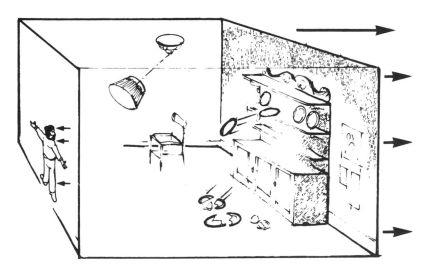

The Accelerating Room

of magic. It seemed at first sight as if 'the same laws of mechanics' are *not* 'valid for *all* frames of reference' as Einstein had maintained. Should Relativity then be confined to simple cases of steady speed only? No scientist with Einstein's total view could possibly rest content with such a verdict and we know he began work on these problems almost immediately after the Special Theory for he published his first tentative paper on General Relativity within two years. Unlike the Special Theory this was a real one-man creation and, although still controversial, it offers the possibility of understanding *space* and *gravity* in a completely new way.

When any object accelerates it is subject to a force—larger for more massive bodies and smaller for lighter ones. Sitting passengers in a

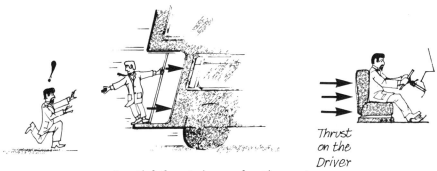

Thrust on the Driver

Inertial thrusts in accelerating systems

bus, for example, feel the thrust on their seat and backs pushing them on. Standing passengers can only be urged forward with the acceleration of the bus where they are in contact with the floor or the hand-rail. Any object without such a force acting on it cannot keep up with the acceleration and will be left behind. Head-rests for car drivers are sometimes provided so that, if the acceleration is very great, there will not be an uncomfortable wrench in the neck as the body and shoulders are pushed forward leaving the head unsupported. Such an effect—the inability of an object to accelerate without force—is called 'inertia' and the amount of force required depends on this 'inertial mass' of the body as well as on its acceleration. The equation for this force was given by Isaac Newton as

$$\text{FORCE} = [\text{Inertial Mass}] \times [\text{Acceleration}]$$

Now under the force of gravity Galileo had observed that all objects fall with exactly the same acceleration. Newton had also verified this point with his hollow pendulum experiments so that it was fair to assume that the force of gravity was proportional to the inertial mass of a body. Indeed we commonly equate the 'weight' of an object with its 'mass' although the result is really quite surprising. *There is no other force in Nature, apart from gravity, which behaves in this way*, being larger for greater masses and less for smaller ones, so that the acceleration of all is exactly the same—*except the inertial forces which arise inside an accelerating system.* Two hundred and fifty years after Newton it was this simple point which was picked up by Einstein and given significance as the cornerstone of his Theory of General Relativity. He called it the 'Equivalence Principle'.

> This experience, of the equal falling of all bodies in the gravitational field, is one of the most universal which the observation of nature has yielded; but in spite of that the law has not found any place in the foundations of our edifice of the physical world. (Einstein. 'On the Influence of Gravitation on the propagation of Light', *Ann Phys*, 1911.)

Just as in Special Relativity Einstein had maintained that the measurements made by two relatively moving observers were exactly equivalent and could never determine which one was at

Both observers, A and B, can feel a 'weight' in their hands and believe that they are in a gravitational field

rest, so now he set out to prove that an accelerating observer in empty space could be equally convinced that he was at rest on earth under the influence of gravity. For him everything would fall to the ground (ie, the floor) with the same acceleration exactly as on earth. If he held up an object it would 'weigh' on his hand, from which it acquires the necessary inertial force to accelerate just as it would in a gravitational field. In this sense acceleration, too, could be seen to be 'relative' when it was taken in conjunction with gravity.

This assumption of exact physical equivalence makes it impossible for us to speak of the absolute acceleration of the system of reference, just as the usual theory of relativity forbids us to talk of the absolute velocity of a system; and it makes the equal falling of all bodies in a gravitational field seem a matter of course. (Einstein, *op cit*.)

This Equivalence Principle can also be illustrated in the opposite way by letting the earth-bound observer fall freely (accelerate downwards) in the pull of gravity. He would then find objects

144

Falling freely towards the Earth

'Stationary in Space'

Both observers feel weightless and devoid of gravity

apparently floating beside him since they all accelerate at exactly the same rate and he might well assume that he was weightless and at rest at a remote point in space. Something of this sort can be felt in an old-fashioned lift as it starts its descent. The sickening 'lurch in the stomach' as it accelerates downwards is due to a momentary reduction in the apparent force of gravity which leaves us with so little weight that the normal reassuring pressure of the floor on our feet and the internal organs on our stomach is much less than we are used to. True weightlessness would result if the lift cable broke altogether (assuming there were no safety devices)—not an experiment that one would advocate!

GRAVITY AND TIME

In the paper which has just been quoted Einstein examined some of the other consequences of this Equivalence Principle. He predicted two new effects that gravity should have upon light, so linking together these two distinct phenomena in a way which would have delighted Michael Faraday. First he established a 'time dilation' due to gravity which he argued in the following way.

Imagine a 'clock'—or any natural timekeeper—floating freely 'at the top' of a box in space which is then accelerated sharply 'up-wards'. When the floor hits the clock there may be considerable relative speed between the two (depending on the distance from ceiling to floor) and Special Relativity would predict a slowing down of the clock relative to an observer at floor level. Even a beam of

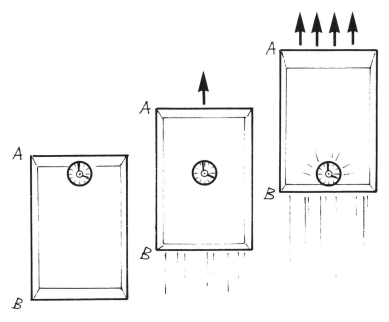

Clock is at rest at A but moving fast when it reaches B due to the upward acceleration

light travelling at its usual lightning speed from ceiling to floor could take sufficient time for the floor to have speeded up appreciably during its transit if the acceleration were great enough. Because light waves are electromagnetic oscillations of a definite frequency they, too, are timekeepers and an observer on the floor should see the light slowed down, vibrating at a slower frequency, and there-fore of a slightly different colour, as though it had been shifted towards the red end of the spectrum. At this point Einstein used his Equivalence Principle to predict that exactly the same effect should take place within an equivalent gravitational field without any acceleration. He could calculate, for example, the precise amount of 'red shift' which should be observed in light which originates within the intense gravitational pull on the surface of the sun. Since the spectrum of sunlight is crossed by numbers of well-defined dark lines corresponding to known elements it was only necessary to compare the position of these lines with those produced on Earth

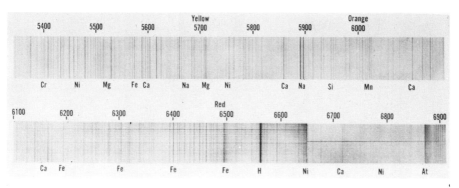

Part of the spectrum of the Sun showing the fire black 'Fraunhofer' lines

to subject Relativity to its first experimental test. In fact a 'red-shift' in the Sun's spectrum had already been detected, as Einstein was aware, and although it was of just the right magnitude to verify his theory it had been attributed, quite credibly, to the high pressures existing on the surface of the Sun. More recently larger 'Einstein red-shifts' have been found in the spectra of the massive 'white dwarf' type of star whose immense force of gravity can make the oscillations of a light ray considerably more sluggish.

About ten years ago the reverse effect was also detected by an astonishingly accurate measurement of the colour of a light which was dropped from a high tower. Such a freely falling object should be equivalent to a 'weightless' system in space and will therefore be 'bluer' than such light is normally in the Earth's gravitational field. A minute change in frequency of exactly the right amount was successfully recorded by using very delicate and complicated instruments.

GRAVITY AND LIGHT

But in 1911, when Einstein published this paper, he had a different experimental test in mind. In an accelerating system the steadily increasing speed would result in clocks running slower *at different levels* within its space. A similar result could be expected under the influence of gravity so that an observer measuring the speed of light would find it faster higher in the field and, using the same clock, slower when measured nearer to the source of gravity. Such an effect should bend a beam of light which was travelling past a massive object like the Sun in the same way as a troop of marching

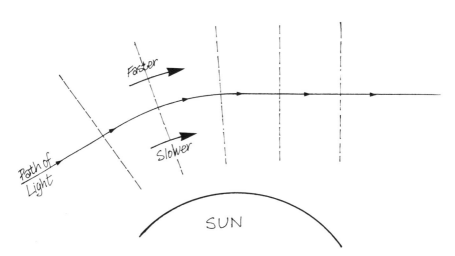

Bending of a ray of light in the Sun's gravitational field

soldiers swing round when the outer file marches faster than those on the inside. Does the light coming from distant stars beyond the Sun really do this? Such stars would be hidden in the glare of direct sunlight at most times and Einstein suggested that observations should be taken during a total eclipse of the Sun. He was eager to test out his new theory by direct measurement.

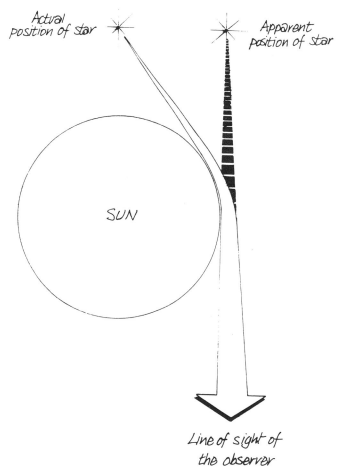

Displacement of a star by the bending of its light

149

A ray of light going past the sun would accordingly undergo deflection to the amount of . . . 0·83 second of arc. . . . As the fixed stars in the parts of the sky near the sun are visible during total eclipse of the sun, this consequence of the theory may be compared with experience. . . . It would be a most desirable thing if astronomers would take up the question here raised. (*op cit.*)

Shortly afterwards an expedition did set out to take measurements during a total eclipse of the Sun which was visible in Russia, but they were caught by the outbreak of World War I and interned. However, Einstein's international reputation rose above national hostilities—as it did again thirty years later—and in 1919 it was British expeditions that went to Brazil and West Africa to verify the predictions of this German professor. In spite of poor weather conditions a small displacement of the stars beyond the Sun was successfully established. The newspapers of the world took notice and Albert Einstein became a celebrity overnight and the scientific prodigy of the age.

13
General Relativity.
The Curving of Space

Meanwhile Einstein had gradually achieved a startlingly new approach to the nature of the 'field of gravity' around an object. He had never suggested that some instantaneous force leapt out of the Sun to pull at beams of light and other masses as Newton's theory of gravitation had always seemed to imply. Even in the Special Theory he had established that 'simultaneity' was only relative, so such an instantaneous action-at-a-distance was quite unthinkable. He believed that the field of influence of gravity extended throughout space as Faraday had visualised, but its effects were proving far more dramatic than he could have ever foreseen. It was clear that measuring rods would contract to different lengths at different levels in the field, time itself would run slower at some places than at others and rays of light would not travel in straight lines wherever strong gravity holds sway. In such a distorted region of space-time it is fair to ask 'What *is* a straight line?' Since neither rulers nor beams of light are reliable it becomes a very fundamental question.

CURVED GEOMETRY

Long ago, in the geometry of Euclid, a straight line had been defined as 'the shortest distance between two points' but the effects of strong gravity are so curious that this definition is no longer of much help. In fact geometry had never been a *science*, in spite of its misleading title; it was simply an elegant 'closed' mathematical system which started from its own arbitrary principles and arrived

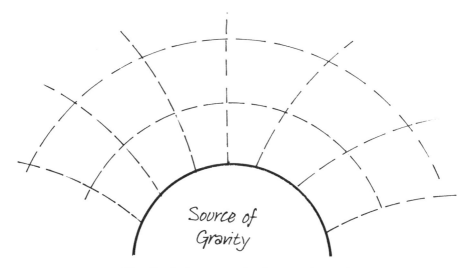

'Bending' of a network of measuring rods

at conclusions about shapes, lengths, and angles by strictly logical arguments. Yet it all seemed to describe the world around us so completely that it had stood unchallenged for nearly two thousand years. In the early nineteenth century the great German mathematician K. F. Gauss had developed a new sort of geometry, drawn on curved surfaces, where 'straight lines' were parts of great circles like the lines of longitude on the Earth. In this geometry many of Euclid's 'self-evident' axioms no longer hold true—parallel lines do meet and squares have angles of more or less than 90°. Gauss treated this as more than just a fascinating mathematical exercise; he was prepared to challenge Euclid's 'flat' geometry against real measurements. He launched an attempt to detect actual 'curvature' in a huge triangle formed by the peaks of three high local mountains by measuring the angles at its corners. Within the accuracy that he could obtain these angles added up to 180°, as every schoolchild would expect, but it still stands out as a remarkable feat of creative scepticism. Gauss' pupil, Riemann, later developed the curved, non-Euclidean, geometry of *four dimensions* which Einstein was now to seize upon in order to resolve the dilemma of gravity.

DISTORTING THE FOUR DIMENSIONS

The results that Einstein had obtained using his Equivalence Principle, startling and new though they seemed, were still incomplete. He was seeking, as always, for a universal mathematical solution and

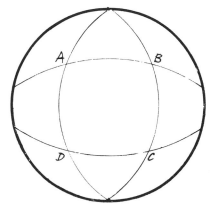

A 'square' ABCD whose angles are all greater than 90°

not primarily for isolated effects of the force of gravity on light. Even those for whom the 'delights of mathematics' is a contradiction in terms can appreciate something of his ambition and dilemma! It was not satisfying to have observers at rest (or moving steadily) describing the laws of Nature quite differently from those who were accelerating (or in a gravitational field) as though they each inhabited different universes. Einstein was looking for a mathematical transformation, similar perhaps to the famous Lorenz transformations, that could link together the totality of *all* experiences. For some years he searched in vain until he was introduced to

Riemann's 'curved' geometry. Using this very flexible form he was able to construct much more general equations in which the 'field of gravity' turned out to be, quite precisely, the amount of actual curvature in that region of space-time. Any object at rest (or moving steadily) far away from the attracting stars would be in a region of almost 'flat' pseudo-Euclidean space-time, but acceleration or massive objects embedded in this all-embracing continuum seemed to distort it like weights on a stretched rubber sheet. Gravity had become a new kind of geometry!

Against this background of continually varying scales and curvature Einstein's equations showed that projectiles and rays of light did follow 'straight lines', but that these were four-dimensional geodesics (similar to 'great circles' on the surface of the Earth). Far from the pull of gravity such geodesics would seem recognisably straight even to our simple three-dimensional perceptions; but where gravity bends the continuum in which such objects travel the geo-

A model of the gravitational field embracing the Sun, Earth and Moon. The four dimensions of space-time have been reduced to the two dimensions of a rubber sheet so that the curvature and distortion produced by an embedded mass can be clearly seen

desics would also bend; and curvature in the fourth dimension, time, would involve changes in speed and energy as well. The geometry of each region now became the dominant factor in its local traffic laws and the established way of looking at gravity was completely reversed. The individual masses that Newton had seen influencing each other across the blank neutrality of a vacuum were now reduced to mere puckers or depressions in the all-powerful medium which alone guides the motion of a particle. It is reminiscent of the way in which a golf ball is swung off-course by the 'lie of the land'. Most holes are a little 'funnel-shaped' at the top and if the ball is aimed true it will roll straight and sure into the hole accelerating a little where the ground slopes down; just as a planet, if travelling straight towards the Sun, would fall into its surface. But if, as so often happens, the ball arrives at the hole slightly off centre the dip around the hole will swing the path of the ball which will then obstinately refused to drop in; this is equivalent to the history of a comet rushing towards our Sun, swinging round it and careering off again deep into space.

'OPEN VERDICT'

Einstein's equations proved their worth in three distinct ways. First, they showed how the old laws of conservation of energy and momentum must arise, quite inevitably, as objects slide 'down' the geodetic contours of this four-dimensional maze. Second, they yielded, to a first approximation, exactly the same results for planetary motion as Newton's laws had done so successfully in the preceding centuries. In the third place, they scored a spectacular triumph in that region where the force of solar gravity (or the curvature of space-time) was at its greatest. Back in 1859 it had first been observed that the orbit of the innermost planet, Mercury, was not exactly a closed ellipse as the Newtonian laws would predict and that its orbit very slowly revolved around the Sun in a kind of rosette shape, called a perihelion. This discrepancy amounted to no more

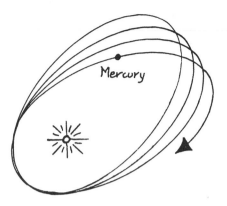

Path of Mercury (much exaggerated)

than 43 seconds of arc (60 seconds = 1 minute. 60 minutes = 1° of angle) during a whole century but Einstein discovered, to his delight, that he could predict just such an amount of twist using his new equations. He concluded his famous paper on 'The foundation of the General Theory of Relativity' with these words:

> These equations give Newton's laws as a first approximation and lead to an explanation of the perihelion of the planet Mercury. . . . These facts must, in my opinion, be taken as convincing proof of the correctness of the theory. (*Ann der Phys*, 1916.)

In spite of the same title 'Relativity' this theory has a very different complexion from the Special Theory. That was a tool, simple to use and easy to interpret; but the General Theory is speculative, revolutionary, and mathematically very difficult to handle. It inspires delight and controversy in physicists and philosophers alike, being hard to test by direct observation and almost equally hard to defend against logical attack. Einstein erected it, not to explain how effects follow causes, but to stand like an illuminating guide to the nature of space-time. He claimed only that his theory was 'psycho-

Polythene model (above) to show the curvature of a two-dimensional field around a massive object. Translucent Variation on Spheric Theme—Gabo (below). At almost exactly the same time constructivist sculptors were trying to achieve '. . . Shapes which represent the fluctuating play of tensions and forces'

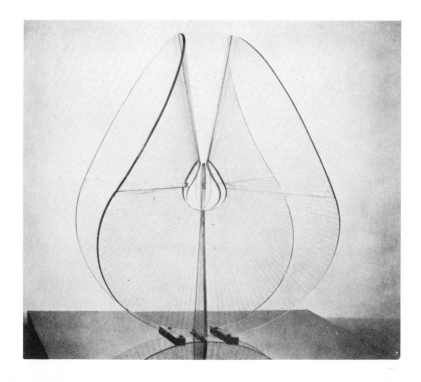

logically natural' but to his followers it has seemed to have a kind of cosmic beauty—mysterious, elusive, but endlessly inspiring. His friend, the distinguished physicist Max Born, who knew him at this time, wrote of the General Theory :

> The foundation of general relativity appeared to me then, and it still does, the greatest feat of human thinking about Nature, the most amazing combination of philosophical penetration, physical intuition, and mathematical skill. But its connections with experience were slender. It appealed to me like a great work of art, to be enjoyed and admired from a distance. (Lecture 'Physics and Relativity', 1955.)

IF SPACE-TIME IS RELATIVE HOW DO YOU EXPLAIN NEWTON'S BUCKET ?

The General Theory inspired feelings of unease as well as admiration for it had quite reversed the teaching of antiquity. It had reduced the status of solid matter to no more than a local distortion in the potent but intangible 'space-field'. Worse than that, it had finally rejected the notion of Absolute Space which had formed such a sure and unchanging background like a familiar map pegged down by the constellations of the fixed stars. Now both gravity and acceleration had been shown so to distort space-time that it lost its immutable character and became no more than a subjective experience of the observer, stressed and curved by his presence and movement through it for all the world as though it were a vast ocean of thick treacle ! The twentieth century has not been kind to any fixed standards of opinion or morality.

Long ago Isaac Newton had argued the existence of Absolute Space by his imaginary experiment with a rotating pail of water. If the surface of the water was flat this was taken as clear proof that the pail was at rest in relation to universal space, if curved it was in detectable and absolute movement. Ernst Mach, that great opponent of the absolute, attacked this argument, observing that it was the 'fixed stars' that Newton was using as signposts to an imaginary Absolute Space. Since all we can measure is *relative* motion, the spinning of the bucket against the background of the stars is indistinguishable from the spinning of the stars round a stationary bucket. Would it not be more logical to ascribe the 'dishing' of the surface of the water to the attraction of *real* stars rather than to some hypothetical Absolute Space?

Could remote rotating stars really be affecting every object on the earth with so much power? A cast-iron flywheel can be shattered by spinning it above a certain critical speed; is this due to 'the action of Absolute Space' or to a kind of gravitational influence from billions of tiny pin-points of light? Neither seems immediately likely. Bertrand Russell, like most other contemporary philosophers, rejected Mach's ideas out of hand:

> It is urged that for 'absolute rotation', we may substitute 'rotation relative to the fixed stars'. This is formally correct, but the influence attributed to the fixed stars savours of astrology, and is scientifically incredible. (1927.)

There is a side-show at the fair which demonstrates intuitively a connection between rotation and gravity. It is called the 'Rotor' and consists of a cylindrical chamber which can be spun round so fast that the victims within it are pinned firmly against the wall by centrifugal force while the floor beneath their feet is removed. My children found the experience fascinating and reported that they always believed the child on the opposite side was higher up than they were. I was sufficiently curious about this to submit to the ordeal myself and found, as they had done, that the powerful centrifugal force which pushed one outward against the side was so identified with gravity that it gave the clear impression of lying peacefully on one's *back* looking *upwards* at the people on the other side while the spectators whirled across one's field of view like a rainbow of merged faces. I emerged from the experience a little giddy but understanding Mach's Principle more clearly than ever before. For a short time the relatively revolving watchers, like the 'fixed' stars, had seemed to give me a new horizon of gravity.

THE UNIVERSAL GRIP OF GRAVITY AND THE INERTIA IT PRODUCES

In recent years Einstein's equations have been re-examined from this point of view and it has been shown that they do predict a gravitation pull from a revolving universe of stars of exactly the magnitude of this inexplicable centrifugal force. Rotation is only relative to real objects and an observer who sees the stars spinning round him can attribute his vertigo directly to their influence! Not just rotation but all kinds of acceleration can be interpreted in this way. When a space-rocket is launched the astronauts can see the stars

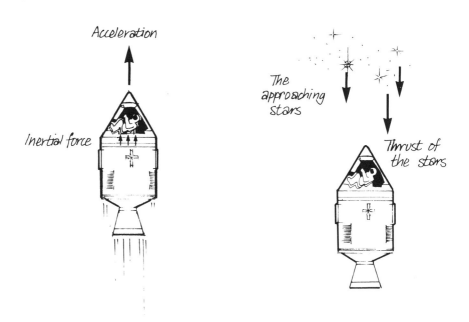

apparently rushing towards them and can understand the crushing inertial force they feel—it is simply the gravitational thrust of the approaching stars!

It is an immense conception that all the humdrum jolts and jars which we experience should be due to the controlling presence of these remote stars. Whenever a car takes a corner or a bus brakes suddenly it is their distant gravity which we feel as our inertia. The total mass of all those far-away worlds must be enormous and if they were to move, collapse and destruction might well follow on our earth. Yet, in the sense of Relativity every time one turns round it makes the whole firmament revolve—no wonder the blood rushes to one's head! The force of Newtonian gravity, which is responsible for ordinary weight, acts between any two bodies whether they are moving or at rest; but Einstein's equations predict an additional force that operates only when there is relative rotation or acceleration. It is a little like the wake of a *moving* boat or the magnetic field of a *moving* charge, and it gives rise to all our inertial mass. No wonder that gravity and inertia always resemble each other so closely—both are gravitational effects and the Equivalence Principle is explained at last. By this new interpretation the force of gravity unites every object in its universal web, even the humblest pebble on the beach is constrained in every movement by the sway of stars which are millions of light-years away just as they, too, must be influenced a little by it! Like so many other twentieth-century ideas it appears both democratic and yet totalitarian for it combines all its members in one vast system where each contributes a little towards the very forces which bind the whole society so rigidly together.

In this and many other ways Relativity continues to stimulate new ideas and yet it must be admitted that it has still not gained a complete measure of acceptance. There seem to be three distinct reasons for this reluctance on the part of physicists:

(1) 'Curvature of space-time' seems such an abstract, geometrical concept for describing the reality of gravity that some have pre-

ferred ideas more clearly based on the measurements made—but every rival theory has proved to have its own drawbacks.

(2) Einstein's equations are so extremely difficult to handle or solve that there are many problems for which they have proved to be of little help.

(3) Gravitation is a very weak effect compared to electromagnetism and very few experimental tests can be devised that will yield results large enough to confirm or refute a theory which approximates so closely to that of Newton for all large-scale effects.

14
Gravity Waves and the Unstable Universe

In the last few years there has been an exciting resurgence of interest in gravity. This may be partly a result of man's own adventures into space and partly due to the astonishing results of new cosmologies but it certainly has stimulated research into new phenomena using novel techniques.

Plans are going ahead to instal gyroscopic compasses in earth satellites to be monitored for twist as they are carried round the Earth. Careful observations of dense stars are being carried out to detect 'black holes' where the pull of gravity might be so strong that no light whatever could reach astronomers on Earth. Experiments have also been in progress to detect a new kind of wave which might be radiated to us from any distant star which suffered a sudden change of shape.

RIPPLES IN 'SPACE-TIME'

In his first paper on Relativity Einstein had predicted the existence of these gravitational waves from the mathematical form of his equations. He described them as changes in the curvature of space-time which would be propagated through the universe at the speed of light vibrating any masses that they met on their way as ripples in the water wobble a floating cork when they pass by. This is so reminiscent of the prediction of radio waves by James Clerk Maxwell that another Hertz might have been expected to confirm their existence within the next few years. History did not so easily repeat itself because, as was known, such gravitational waves could be expected to be extremely weak. A spinning rod one metre long, for example, might rotate at speed for billions of years before it radiated even a tiny fraction of its energy and the chances of detecting so feeble a wave seemed vanishingly small. However, there was the possibility that collapsing stars or some other violent catastrophe somewhere in the universe could be generating much stronger waves which were shaking us imperceptibly all the time in the rhythm of their unknown frequency. This challenge to experiment was taken up and some results were published in 1971.

Gravitational waves are 'quadripole': this means that they would oscillate a pair of crossed rods placed at right angles to their path in two opposite senses at the same time.

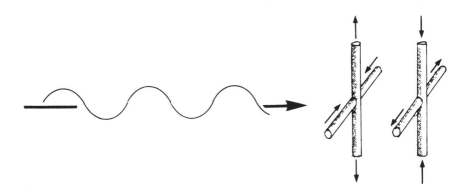

Gravitational waves causing quadripole oscillations

Any large object might be set into resonance by such waves if its own natural frequency of vibration was the same as that of the wave, in the same way as an empty bottle resounds if the right note is hummed at it. In theory all that was needed to detect these waves was some large absorbing object which was freely suspended within a vacuum chamber and carefully cushioned against all external jars and jolts. Then, if it were being subjected to gravitational oscillations of its own known frequency, the resulting internal vibrations might be picked up by sufficiently sensitive instruments. This is the type of apparatus that was used recently in the University of Maryland. The object was a large aluminium cylinder five feet long, three feet in diameter, weighing more than three tons and

The enormous aluminium cylinder used to detect gravitational waves

fitted with a belt of delicate crystal detectors round its centre. These crystals react to changes in pressure by producing a small electric potential across their faces which can then be accurately measured. In fact the sensitivity of the apparatus was such that changes in size of no more than one-hundredth of the diameter of a single atom could be recorded—a minute deformation in a cylinder five feet long! This size of receiving object was chosen because its resonant frequency, about two to three octaves above middle C (approximately one thousand cycles per second), would be the same as the frequency of wave that a star the size of our Sun might generate if it suffered gravitational collapse.

STRANGE RESULTS OF THE EXPERIMENT

At first sight the figures obtained from these experiments differed very little from completely random fluctuations in shape. They had first to be compared with the vibrations of another similar cylinder situated eighty miles away so as to eliminate disturbances due to local Earth movements, and then the orientation of the cylinder with respect to the Sun and the stars had to be considered. Twice during the 'sidereal day' (23 hours 56 minutes) when the cylinder was lying broadside on to the direction of the galactic centre there was a quite definite peak in amplitude of its response. These are the times when gravitation waves from deep within the Milky Way would impinge on the cylinder with maximum effect as it is carried round on the surface of the Earth. The Sun itself was not the source of this new sort of cosmic radiation.

Much more needs to be discovered about these waves. So far they have only been shown to be generally quadripole in nature, as Einstein predicted, but a much larger detector is needed for better absorption of the waves and further experiments using the Moon itself are planned for the future. Astronauts on the Apollo 17 mission are expected to leave apparatus on the lunar surface which will 'listen' for oscillations of this still, quiet sphere as she orbits

166

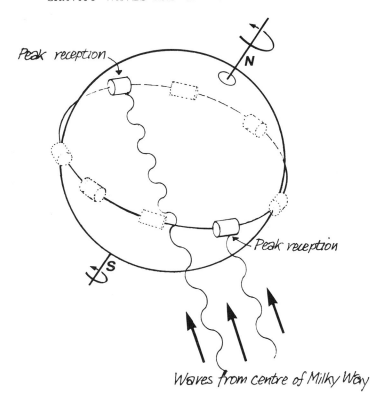

Waves from centre of Milky Way

our more turbulent Earth. The source of these waves is a total mystery. They have turned out to be far stronger than astronomers could have foreseen, although still barely detectable, for more than a thousand suns would have to collapse into nothingness every year to radiate so vast an amount of this elusive form of energy. All the light and radio waves emitted from our whole galaxy do not equal its volume and the only explanation offered so far seems to touch the borderlines of pure fantasy.

The origin of the observed gravitational radiation has not been determined, only the direction of its arrival. It is conceivable that the source might be an unusual object such as a pulsating neutron star very much closer than the galactic centre. It is also conceivable that the mass at the galactic centre is acting as a giant lens, focusing gravitational radiation from an earlier epoch of the universe. Since gravitational radiation is not appreciably absorbed by matter, it should have been accumulating since the beginning of time. The relatively large intensity apparently being observed may be telling us when time began. (J. Weber, 'Detection of Gravitational Waves', *Scientific American*, May, 1971.)

RELATIVITY IN COSMOLOGY

This strange conclusion to a report on sober practical research in a conventional laboratory is a reminder of the contemporary revolution in cosmology—the theories about the evolution of the Universe. This advance, too, is bound up with the emergence of General Relativity. Einstein himself lost no time in extending his ideas on the curvature of space-time to cover the entire cosmos. There was no reason to suppose that any such curvature existed in regions far

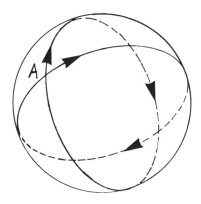

Two-dimensional curvature, on the surface of a sphere, causes all lines radiating from A to encircle the sphere and return to A

removed from the gravitational pull of stars, so that beams of light, radio waves, and even errant stars and meteors might be expected to travel through such spaces in endless straight lines to disappear for-ever into the depths of 'infinite space'. Einstein could not accept such an impossibly uneconomic state of affairs which would eventu-ally lead to the running down of the whole Universe by continual loss of radiation and matter. On paper, at least, he could resolve this difficulty by introducing a little arbitrary curvature into the whole of space-time. The two-dimensional analogy of this procedure is quite easy to visualise. On a flat sheet of paper straight lines drawn from a point could be extended outwards for ever (mathematicians would describe them as of 'infinite length'). On the curved surface of a sphere such lines would trace out complete circles and finally reappear at their starting point from behind the back of the sphere. In the four dimensions of space-time much the same effect can be achieved. Although there is no sudden edge or sharp boundary to such a Universe its total space is *not* infinite, and no matter or energy which is flung out from one point in its space can ever entirely escape. Using his powerful equations Einstein proved that such curvature would have to be very slight indeed and smaller in proportion as the density of matter in the Universe was less.

This work began as little more than an abstract mathematical exercise prompted by Einstein's personal preference for a kind of cosmic conservation. Very soon other mathematicians were juggling with the complex equations of Relativity and concocting other models of the Universe which would conform to them; however, in 1922 it was shown that all such 'curved universes' suffered from the same serious drawback—they were essentially unstable and would, if left to themselves, either contract into a pin-point or expand for ever! It seemed, for a while, that the whole theory must be at fault. If the background of space-time itself was steadily changing its curvature then every star within it would either seem to be reced-ing from every other star at an ever-increasing rate, or else they would all be heading towards a final cosmic pile-up! It was as if the stars were blobs painted on a toy balloon that was either being blown up or deflated and so altering the distance between the stars. Far from discrediting these first efforts at relativistic cosmology, observation was about to confirm just such a strange unlooked-for effect in the next few years.

15
Beyond Our Island Universe

Astronomy had made slow but steady progress throughout the nineteenth century. The distances of the nearer stars had been measured accurately, for the first time, and great catalogues of the stars and the nebulae had been drawn up, but Kant's daring theory that these faintly glowing nebulae were great systems of stars, 'island universes' similar to our own Milky Way, was still no more than sheer speculation. Their distances defied all measurement. It had become clear, however, that the nebulae fell into two distinct categories. There were those, like the Orion nebula, that had a cloudy, diffuse shape, were always to be found within the path of our Milky Way and gave a spectrum like that of white-hot gas. The others, like the great Andromeda nebula, seemed to avoid the disc of the Milky Way, they gave a spectrum which was crossed by fine dark lines like that of the Sun and the stars and, when viewed through a telescope, stood revealed in beautiful spiral and elliptical shapes poised in space at different inclinations to our view. Only these nebulae could fit the rôle of vast external systems of stars far out in space, but there was still no shred of solid evidence to support Kant's theory.

STARS ARE SEEN IN THE NEBULAE

In 1885 an extraordinary event was observed within the great Andromeda nebula—a brilliant spot of light suddenly blazed out shining with a radiance which was comparable to that of the whole nebula. Then it faded away. Immediately the whole controversy as to the nature of these nebulae began again. Stars which suddenly flared up and then died away again were well known within our Milky

Sa NGC 4594

SBa NGC 2859

Sb NGC 2841

SBb NGC 5850

Sc NGC 5457 (M101)

SBc NGC 7479

Spiral nebulae

Way, but if this explosion within the Andromeda nebula was only such a 'nova' then the whole system had to be both much nearer and much smaller than Kant's prediction. Alternatively this remote new 'star' might be in quite a different class, a rare 'supernova' like that which Chinese astronomers had recorded during 1054 in the constellation of the Crab which could be seen shining brightly in broad daylight. The problem seemed insoluble though there were valuable clues to be found in the southern skies where, unfortunately, there were no telescopes to observe them—the neighbouring galaxies are definitely not laid out to suit the convenience of earthly astronomers!

Down below the equator the constellations are unfamiliar. On the whole they seem less striking than those of the north but in one respect they do surpass our night sky for they contain the two most clearly visible of all the nebulae. These small patches of light in the sky were first reported by the daring Portuguese adventurers of the fourteenth century as they sailed towards the Cape of Good Hope. '. . . certain shining white clouds,' they wrote, 'here and there among the stars, like unto them which are seen in the tract of heaven called Lactea Via, that is the milk white way.' Later when Magellan made his famous voyage round the world he, too, reported seeing these 'shining clouds' which form an equilateral triangle in the sky with the vacant point that is the southern pole; since those days these two neighbouring nebulae have been called the 'Magellanic Clouds' in honour of this great explorer. Unfortunately they do not have the usual spectacular spiral shape but belong to a rarer, more shapeless class of nebulae. However, the first good telescope to be erected in the southern hemisphere was rewarded by the sight of millions of separate stars which together give the Magellanic Clouds their soft glow. Many photographs of these nebulae were made high up in the clear skies of Peru and sent back to Cambridge, Massachusetts, for examination and there, in 1908, a discovery was announced which set the stage for a revolution in practical astronomy.

MEASURING THE DISTANCE OF THE NEBULAE

The small Magellanic Cloud had been found to contain a number of strange pulsating stars that waxed and waned in a very regular way. Some took two days or less to change from dim to bright and back to dim again, some took longer than a month, but in every case the rhythm never changed. Such variable stars had already been noticed within our own galaxy: even our familiar Pole Star is such a 'Cepheid variable'. It has a period of almost four days during which its brightness changes only slightly but always by exactly the same amount and to the same monotonous rhythm. Within the Magellanic Cloud it was discovered that a simple mathematical law connected the brightness of such a star with the length of its period of fluctuation. It was as though the larger stars expanded and contracted more slowly and cumbrously than the smaller, fainter stars in the same way as larger tuning forks oscillate more slowly than do smaller ones and so give a lower note. This had not been at all apparent in the nearer Cepheids of the Milky Way although these are so much clearer and brighter. 'Local' stars may vary in their distance from us by more than a thousandfold and it is always very difficult to distinguish between a small star that seems very bright because it is so near and one which is intrinsically much brighter but lies at a greater distance. Within the Magellanic Cloud, however, *all* the stars are at an equally great distance from us and one that *seems* brighter than another can be assumed really to be so. The 'Period-Luminosity Law' for these slowly pulsing giant stars could be established by direct observation.

We can only guess at the mechanism by which a whole star, thousands of times brighter than the Sun, can wax and wane to its own steady beat throughout the centuries but it was recognised very quickly that these Cepheid variables offered a wonderful opportunity for measuring astronomical distances. The calculation would go as follows:

(1) Time the rhythm of the star in question.

(2) Use the Period-Luminosity graph to find out its intrinsic brightness.

(3) Measure the peak of *apparent* brightness of the star (by examining the density of its image on the photographic negative).

(4) Since distance dims all stars according to the well-known 'Inverse Square Law' a comparison of (2) and (3) will give its distance from us.

It only remains to establish the scale of the graph by measuring the actual distances of a few neighbouring Cepheids. This turned out to be a fairly difficult task since none of them are very near; the Pole Star, for example, is really two thousand times brighter than our Sun but over four hundred light-years away. However, by 1917 the first calculations were complete and this new yard-stick to the

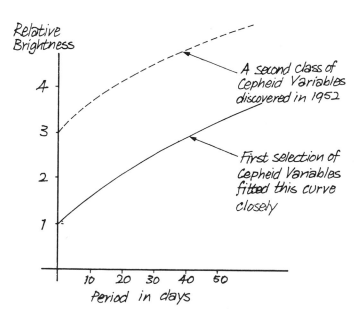

stars lay ready for use at about the same time as the new 100-inch telescope on Mount Wilson came into operation. It was a powerful combination which was to push back the frontiers of astronomy to an inconceivable depth.

Soon the first good photographs of the great Andromeda nebula were taken, and careful examination of the negatives showed a bewildering wealth of very faint stars. In 1922 the first Cepheid variables were identified although, at this distance, the brightest of these were about one hundred thousand times fainter than the faintest star which is visible to the naked eye! Within two years eleven more Cepheids had been confirmed in this great spiral nebula—enough to make the first reliable calculation of its distance. The figure was stupendous, one million light-years, or about 6,000,000,000,000,000,000,000 miles away from us, the Sun, and our fellow members of the Milky Way. In the terms of Immanuel Kant it was, in truth, a remote 'island universe'.

A NEW PICTURE OF THE COSMOS

It was Edwin Hubble at the Mount Wilson observatory who made these first calculations of the distances out to the nebulae. He also studied their shapes, their grouping in clusters as well as the brightness of their various types of stars and by 1935 he had developed methods of estimating their distances to a depth of 500 million light-years. This was much more of an achievement than just such a series of measurements might suggest for it took these great depths of space out of the sphere of dreamy speculation into the full glare of scientific inquiry. Many times before throughout history guesses had been made about the nature of the Universe in the dark and distance expanses beyond our own constellation. Now, for the first time Hubble could present a picture of these regions—'The Realm of the Nebulae', he called it—which was informed by observation rather than by imagination but the picture lost nothing of its grandeur.

The Great Andromeda Nebulae

The Earth we inhabit is a member of the solar system—a minor satellite of the Sun. The Sun is a star among the many millions which form the stellar system. The stellar system is a swarm of stars isolated in space. It drifts through the Universe as a swarm of bees drifts through the summer air. From our position somewhere within the system, we look out through the swarm of stars, past the borders, into the Universe beyond.

The Universe is empty, for the most part, but here and there, separated by immense intervals, we find other stellar systems, comparable with our own. They are so remote that, except in the nearest systems, we do not see the individual stars of which they are composed. These . . . nebulae are scattered at average intervals of the order of two million light-years or perhaps two hundred times their mean diameters. The pattern might be represented by tennis balls fifty feet apart.

The order of the mean density of matter in space can also be roughly estimated . . . about one grain of sand per volume of space equal to the size of the earth. (*The Realm of the Nebulae*—Hubble, 1936.)

MEASURING THE SPEED OF THE NEBULAE

At the time that Edwin Hubble was establishing these distances there was another astonishing set of observations concerning the nebulae which was already in existence. It seemed that they were all rushing about at tremendous speeds. Of course it is rarely possible to measure the speeds of stars directly, let alone those of the distant nebulae. If they were to be moving towards us or away from us in our line of sight, they might be expected to grow steadily brighter or fainter while we watch but again, at such astronomical distances, the change would be quite unobservable. However, there is a method of measuring such speeds by the frequency (or colour) of the light emitted by the stars; it is called the Doppler Effect and is common to all waves.

It arises like this—if a source is moving towards us while it is

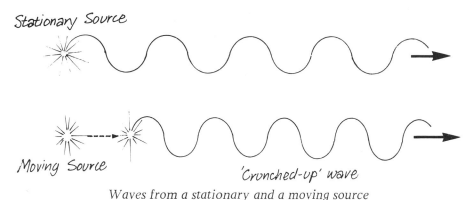

Stationary Source

Moving Source

'Crunched-up' wave

Waves from a stationary and a moving source

sending out waves it is, in effect, almost catching up with its last crest as it emits its next one. The result of this will be a kind of squashing up of the wave so that its wavelength will appear shorter and its frequency correspondingly higher. In sound waves this is a well-known effect which causes the rise in pitch of a car's engine-noise as it approaches us and the drop in pitch as it passes by and moves away. Even the familiar siren of an ambulance or police-car drops by a semi-tone or more as it drives past us and a low-flying aeroplane gives a frightful descending yowl of 'ee-ow' (which needs the lungs of a small boy to do it justice!) due to this same change in frequency. The corresponding effect in the light from a moving star should produce a change in colour—bluer for higher frequencies if they approach us and redder for lower frequencies if they recede. Unfortunately the speed of light is so much greater than the speed of sound that this total colour change is not directly detectable. However, there are some fine black lines in the spectra of stars, whose positions are known very accurately, and small shifts in these lines towards the blue or red end of the spectrum can be measured. When this method was first applied to neighbouring stars in the latter part of the nineteenth century very tiny displacements of these spectral lines were observed giving stellar speeds of approximately 30 miles/sec (about 100,000 miles per hour).

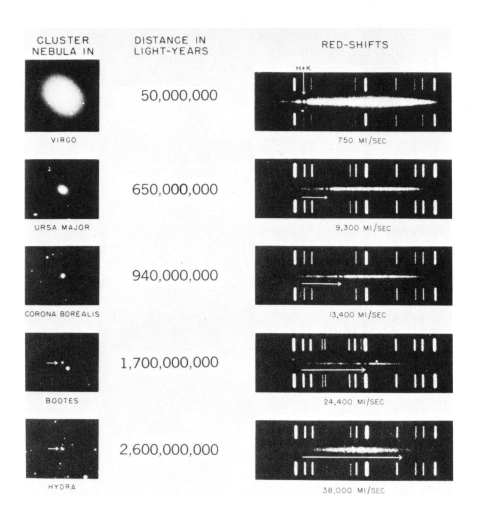

CLUSTER NEBULA IN	DISTANCE IN LIGHT-YEARS	RED-SHIFTS
VIRGO	50,000,000	750 MI/SEC
URSA MAJOR	650,000,000	9,300 MI/SEC
CORONA BOREALIS	940,000,000	13,400 MI/SEC
BOOTES	1,700,000,000	24,400 MI/SEC
HYDRA	2,600,000,000	38,000 MI/SEC

The red-shift of more distant galaxies seems to show that they are moving away from us at ever-increasing speeds

Fast as these speeds may seem it was mere dawdling when compared with the results obtained later for the external galaxies. Speeds
of over 100 miles per second were commonplace and by 1924 the
spectra of the fainter nebulae showed shifts corresponding to speeds
of more than 1,000 miles per second! Such wild and undignified
rushing about in the massive 'island universes' seemed at first very
bewildering. It was strange, too, that almost all these velocities were
directed *away from us* producing 'red-shifts' in the spectra but it did
serve to underline the fact, still under debate at this time, that these
spiral nebulae were far out in space and well removed from the
local forces of gravity that hold our own Milky Way together.

THE EXPANDING UNIVERSE

In 1929 it became apparent that the whole Milky Way was rotating majestically about its centre (far away in the direction of the
constellation Sagittarius) like a great catherine-wheel in space. As
the Sun and all its planets are swept up in this motion we acquire
a velocity of about 150 miles per second which is, at this present
epoch, directed roughly towards the Andromeda nebula. In the
egocentric manner of all observers in space we are bound to ascribe
our own motion to the nebulae we see. When this spurious velocity

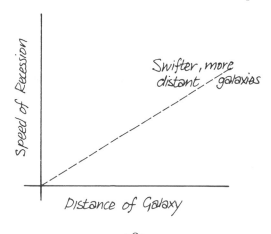

was allowed for every single galaxy could be shown to be receding from us. Now Hubble could use the figures that he had already obtained for the distances of the nebulae to sort out their speeds and show how they fitted into a very simple mathematical law which seemed to operate throughout the whole observable Universe. The velocities with which our neighbouring galaxies move away from us increase quite regularly with their distance from us. The nearer nebulae are almost at rest, those farther away move faster and those farther still are rushing towards the horizon with ever-increasing speeds. The factor by which their speeds increase— 'Hubble's Constant'—appears to be about 20 miles per second faster for every extra million light-years of distance from us. The most distant galaxy whose speed has so far been measured appears to be

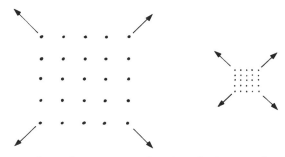

If the separation of an array of points is increased, each point 'sees' its neighbours apparently moving away from it

receding from us at the fantastic velocity of 74,000 miles per second (270 million miles per hour) or about 2/5 of the speed of light itself!

Almost at once it was realised that this strange 'Realm of the Nebulae' was fitting perfectly into the picture of remote space whose curvature had been discussed in the first relativistic cosmologies. If their 'blown-up balloon' model of the Universe was correct then a distance–velocity law exactly like Hubble's would follow naturally. The fleeing nebulae would be carried outwards by the expansion of space-time itself. Observers in other galaxies, too, would see the rest of the Universe retreating from them as if they were the victims of some dreadful cosmic plague! There is nothing sinister in this, it is only due to the stretching of the space between the galaxies, and the Universe would seem exactly the same, with the same red-shifts in the spectra of the nebulae, from every other vantage point. This is the famous 'Cosmological Principle' which relativity has introduced into all modern theories of the Universe—it must look the same overall, to every observer, for it can no more have an absolute centre or edge than space-time itself can be absolute and immutable.

16
The New Cosmologies

At once the question of the 'age of the Universe' and the nature of time itself became a burning issue. This was, in part, a dramatic consequence of Hubble's Law, for if a nebula with twice the speed of another were twice the distance away and one with three times the speed were now three times the distance away it seemed to follow that there had been one special instant, far back in time, when they had all started off from the same place. Was this the beginning of the whole Universe? In the 1930s eminent scientists like Sir James Jeans and Sir Arthur Eddington were writing popular books about the latest developments in physics and 'the Expanding Universe' became a phrase to conjure with. Speculation about a moment of creation left the dreamy regions of metaphysics and religion and became, at last, a serious scientific hypothesis in an age which had already dared analyse and discuss almost every other accepted doctrine.

Time is the ultimate mystery. Nineteenth-century scientists had begun to establish some terrestrial time-scales; Darwin had made sense of the slow progress of animal evolution and the new science of geology had shown the fossil imprints of the passage of time in the sequence of rock layers. The great age of steam-power had stimulated a study of heat exchange and energy transformations and here again the direction of the flow of time stirred the imagination with concepts like 'the running-down of the Universe'. Perhaps prosaic science held the key to the understanding of time itself, whose fascination has held sway since the days of primitive man. Today we are assaulted by predictions on every side; sociologists predict population explosions, conservationists predict total pollution, futuristic science fiction flourishes and there is an extraordinary resurgence of belief in astrology. Eastern religions with their emphasis on cycles of time have captured the imagination of many of the younger 'hippy' generation and this century has also seen the rise of the first new philosophy of time that Great Britain has known for a long time—John Dunne's theory of Serial Time. But it has been within the sciences of astronomy and cosmology that this search has reached its most daring peak.

THE BIG BANG

The first of these modern cosmologies had its origins in the work of the Belgian Abbé Lemaître in 1927 and it is with us still under the simple and striking title of 'The Big-Bang Theory'. It asserted that not only space-time but all the matter of the galaxies has expanded from one explosive Genesis so that the origin of time, space, and substance was one and the same instant. Lemaître wrote simply of the 'primeval atom'—a compact and immense nucleus, composed of all the matter that ever was or ever will be, crowded into a ball no larger than the yearly orbit of our Earth, which exploded long ago like a super atomic bomb throwing out fragments so rich in variety that they provided us with all the different elements that we know today. The expansion that we now observe by the red-shift of the distant nebulae is continuing evidence of the violence of that moment of the Universe's birth. Despite its dramatic flavour this is a serious scientific theory which has been worked out in sufficient mathematical detail to explain how the ejected pieces of primitive matter later condensed into galaxies and clusters of galaxies as we know them today. Natural radioactive elements like uranium and radium are simply overlarge fragments of that primeval atom which are still slowly disintegrating. Also the astonishingly swift atoms which impinge continuously on the upper layers of our atmosphere and are known mysteriously as 'cosmic rays' are explained as energetic

Big Bang

remnants of that first bang which are still speeding away from its ancient eruption at almost inconceivable velocities.

> We picture the primeval atom as filling space which has a very small radius (astronomically speaking). . . . This atom is conceived as having existed for an instant only, in fact, it was unstable and, as soon as it came into being, it was broken into pieces which were again broken in their turn; among these pieces electrons, protons, alpha particles, etc, rushed out. . . . When these pieces became too small, they ceased to break up; certain ones, like uranium, are slowly disintegrating now . . . leaving us a meagre sample of the universal disintegration of the past. (Georges Lemaître, *An Essay on Cosmogony*, 1942.)

Lemaître's theory is cataclysmic, but it is also based on sound relativity theory. It adheres faithfully to the Cosmological Principle and presents no special point in space. Although the moment of his 'big bang' can be calculated from the recession of the galaxies it has no special location that we could make pilgrimage to now. There is no 'site of creation' for *everywhere was where the bang happened* in the instant before the Universe expanded as Einstein, and others, had predicted. At that remote and inconceivable moment the whole of space was included with the whole of matter in the tight knot that was the beginning of everything. These ideas were eloquently championed and elaborated by Eddington, yet the theory has still not won total acceptance and there seem to be two main reasons for this. The first is a kind of aesthetic repugnance to postulating an absolute and miraculous beginning to time which, in Newton's famous words, 'runs always', and especially to one which so clearly demands the fiat of a Creator outside of space and time who would have to set up the unstable primeval atom which then burst instantly into the beginnings of our Universe. Scientists like to keep God out of their equations.

It is the . . . extropolation towards the past which gives real cause to suspect a weakness in the present conceptions of science. The beginning seems to present insuperable difficulties unless we agree to look on it as frankly supernatural. We may have to let it go at that. (Eddington, *The Expanding Universe*, 1932.)

The singular state at the start of the expansion means that certain quantities in our mathematical equations become infinite. Our models of the universe therefore break down, and what we must do is to correct them. There has been a curious reluctance on the part of the cosmologists I mentioned to do this, and they have preferred to identify the singularity in the equations with God. Now I argue that scientifically this is inexcusable, but I should have thought that theologically it is very dangerous, because if God is brought in just to patch up our mathematics then He may become redundant when better mathematics is discovered. (William Bonner, *The Mystery of the Expanding Universe*, 1964.)

The second objection to the 'Big Bang' theory was more precise. Using Hubble's figures for the distances of the nebulae the date of the Bang which began the Universe could be calculated and it turned out to be uncomfortably recent—a mere two thousand million years ago. The only other estimate of the age of the Universe had been made by Sir James Jeans from considerations of the slow burning up of the Sun and the random movements of the stars in space and this was about a thousand times larger. Even geological evidence for the age of the Earth gave it a greater age than Hubble's measurements. Such contradictions were totally unacceptable and this acted as a powerful stimulus for the invention of other theories of the Universe—of which there has been no lack !

All the new theories, except one, have been based on Einstein's General Relativity. Their Universes have been the four dimensions of space and time curved to one pattern or another. Some theories tried to explain the flight of distant nebulae by a new sort of cosmic repulsion which acted like Newtonian gravity in reverse. Other model Universes were possessed of steadily increasing curvature like the Lemaître-Eddington model and some were imagined to oscillate during vast aeons of time passing from phases of expansion to phases of contraction alternately. One theory only rejected the whole complex geometry of 'curved Space-time' and preferred to examine the mystery of time and its measurement in isolation. This was Professor E. A. Milne's very individual contribution during the 1940s.

MILNE'S SCALES OF TIME

Milne started from an examination of how one observer might interpret the information he received by his own time-scale without assuming it to be similar to that used by any other observer. This careful philosophical approach is very similar to that of Einstein in his earlier theory of 'Special Relativity' and it yielded a very interesting result. Different time-scales may never totally synchronise and, in particular, the large-scale time which flows to the rhythm of swinging pendulums and orbiting planets is, Milne believed, basically different from the time-scale of the atom. Of course atoms neither swing nor pulsate in any normal sense but they do 'keep time' in that the colour of the light they radiate has a definite frequency. The intense yellow light of sodium, for example, which is familiar from the depressing street-lights and from the flare of a fire when salt is thrown into it, corresponds to the beating of a sodium atomic 'clock' at the steady rate of 506 million million ticks per second. Milne believed that these two approaches to time yielded quite different scales which do *not* run evenly when compared to each other. They are connected by a logarithmic formula (like the markings on a slide-rule) and this implies that time on the atomic scale would appear to run faster and faster (as observed from the larger scale world) and that large-scale time runs ever more slowly when compared with atomic time.

When this result is applied to astronomy it provides a new explanation of the reddening of light from distant galaxies. The very fact that they are remote—and their distances can be measured in hundreds of millions of light-years—means that the light by which we now observe them was actually emitted long ago when their atoms were much younger and beating more slowly, according to Milne's theory. This would account for a slower frequency and a redder light in the more distant nebulae and for Hubble's Distance- Velocity Law without any need to bring in the Doppler Effect and the almost incredible velocities that it suggests. The farthest galax- ies are at such vast distances that their faint rays of light which we receive on earth are fossil energy from a time so remote that it makes the extinct dinosaurs seem almost contemporary by com- parison! Every time an astronomer looks out into space he is also peering back into time so that the facts of ancient history are, in this sense, observable now, in the present. Milne's theory also pre- dicted an increase in density of the nebulae at great distances. If ancient time ran so slow then we should see galaxies crowded to- gether as we gaze out towards the telescopic horizon and, though there is some uncertainty about the measurements, both optical and radio astronomy have produced evidence that this outward increase in the numbers of galaxies does actually exist.

Milne's work was still incomplete when he died in 1950 and at present it attracts little attention. Nevertheless its novel approach did throw light on several other inexplicable facts of astronomy; the rota- tion of the nebulae, their strange trailing spiral arms and the mysteri- ous increase in speed of our own Earth in her orbit round the Sun which is steadily shortening the 'year' by about one second every cen- tury. For a while the chief attraction of Milne's theory was the way it avoided the awkward problem of dating the Beginning. The large time-scale has no beginning, it is only the atomic time-scale which seems to have commenced in the remote past with infinite slowness and has gradually accelerated up to the present time. The distant galaxies are *not* moving away, only the light from their archaic atomic-clocks seems sluggish to us now. As to when atomic time began, the larger Universe of moving stars can tell us nothing for it runs to a quite different rhythm.

MEASURING THE AGE OF THE EARTH

However, there are atoms here on earth that can reveal a great deal about the age of our world and these are the naturally radioactive

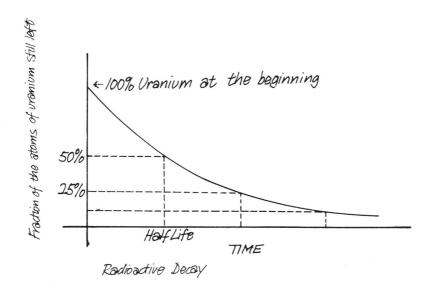

Radioactive Decay

elements: uranium, thorium, and radium. Each of these substances has its own rate of disintegration so steady and inevitable that no variation in temperature or pressure has the slightest effect upon it. We know quite exactly the length of their 'half-life', that is the period of time in which half the atoms in any lump may be expected to break up, emit radiations, and change into atoms of a lighter, more stable element. For uranium this period is 4,500 million years, one of the longest life-spans known. Every atom of uranium is identical, equally unstable and equally near the verge of break-up so that we can never point to any particular atom as the one that will be 'the next to go' nor yet explain after its disintegration why it was that atom which went over the edge while its neighbours, although just as unstable, remained intact. We can only examine the whole collection of over-weight atoms, trembling with excess energy and calculate the *probability* of any one of them disintegrating during the next second. For uranium this is a chance of about one in 200 thousand million—heavy odds against it!—but taken over the vast number of atoms in a single gram of pure uranium it is enough to keep up a steady crackle of more than twelve thousand atomic explosions per second at first, but slowing down with time as the number of uranium atoms left to take part in this radioactive lottery is diminished. This process forms the basis of a powerful new method for measuring the flow of time which can be used to estimate the age of the rocks on the Earth as well as for dating art-treasures, prehistoric skulls, and fossils.

This was a more natural way for physicists to measure time than was the mysterious expansion of the Universe. Time flows always in the direction of 'decay'—tall buildings eventually topple over, mountains are worn away into sand, rivers and glaciers move downwards towards the stable tranquil sea. In the same way uranium atoms, the heaviest in the natural world, suffer 'radioactive decay' gradu-

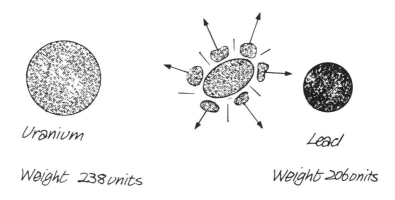

Uranium

Weight 238 units

Lead

Weight 206 units

ally over the centuries and change finally into stable, prosaic lead. Fortunately these lead atoms are very slightly different from common lead, being a lighter 'isotope' which can be distinguished by its smaller atomic weight; so that by careful analysis of the percentage of this sort of lead within natural ores of uranium it is possible to calculate how much uranium there had once been and for how long it has been disintegrating. The figure obtained was 5,000 million years. Once again the duration seemed too long; here was an age for rocks on our insignificant planet that pre-dated the first 'Big Bang' and the beginning of Milne's time-scale for atoms! During the 1940s cosmology seemed to be failing lamentably in reconciling the age of the Universe with the rapid rate of its expansion.

THE 'STEADY STATE' COSMOLOGY

Several efforts were made to extricate cosmology from this head-on collision but none of the alternatives seemed to work. Then, in 1948, a revolutionary new idea took shape. This theory accepted the expansion of the Universe but rejected the whole idea of a distant epoch of creation—the 'singular state' as religion-shy mathematicians tend to call it! If then the galaxies have been moving away *for ever* why do we still have any neighbours? And if matter was not all formed at the beginning how did these galaxies arise? The only possible answer seemed to be that *matter was being created everywhere all the time*. Near us and throughout the whole Universe new atoms are regularly springing into existence; one by one, in isolated spaces, and too rarely to be detected, single atoms of hydrogen are appearing out of nothing! Then, out of this constantly enriched background, gas, stars and galaxies can gradually condense out, gather speed and retreat from view at ever-increasing speeds—but for every distant nebula that disappears over the observable horizon a new one is born in a condensing whirl of the tenuous intergalactic gas. Our overall picture of the Universe never changes. Like photographs of a school taken year after year the individuals may grow

193

up and leave but there are always 'new boys' and younger classes which grow up and take their places.

This 'Steady State Theory' was due to H. Bondi, T, Gold, and F. Hoyle, all of whom wrote enthusiastically of its possibilities and it achieved an immediate impact. Not only did it dispense with the conflict over the age of the Universe but it also re-established the smooth continuous flowing of time. It gave the Universe an immortal status with no beginning and no end and an overall unchanging aspect which seemed aesthetically pleasing—like the attributes of God. However, there were two great drawbacks to the theory. First and foremost the idea of creation of an atom out of nothing seemed utterly at odds with all that science had taught about the conservation of mass. According to Hoyle the 'Steady State Theory' called for no more than one new atom of hydrogen to be created per year in a volume as large as St Paul's Cathedral to compensate for the loss of matter from our vicinity by the expansion of the Universe, but the idea of continuous creation violates a very basic principle in physics. The originators of the theory worked out in detail how all the range of known elements could subsequently be built up from hydrogen by nuclear processes within the stars, but they could do little to cope with the second problem that the theory faced. What makes the Universe expand? Without the mechanism of the first big bang it is hard to see why the nebulae should be speeding outwards at all. The advocates of the theory could only accept it as a familiar 'running down' process like erosion, the smoothing-out of the landscape and the dissipation of heat.

The fundamental assumption of the theory is that the universe presents, on the large scale an unchanging aspect. Since the universe must (on thermodynamic grounds) be expanding, new matter must be continually created in order to keep the density constant. As ageing nebulae drift apart, due to the general motion of expansion, new nebulae are formed in the intergalactic spaces by condensation of newly created matter. (*Cosmology*— H. Bondi, 1952.)

Probably it must seem as if all these rival theories were mere exercises in mathematical imagination, but it is never so in real science. Every theory must be 'testable'. Even if it contains such appealing and emotive words as 'creation' and 'eternity' each hypothesis must lay itself open to both corroboration and falsification or it will become no more than a modern myth. The Steady State Theory, for example, was based on this 'Perfect Cosmological Principle' that the Universe has always looked exactly the way it looks to us now. It followed that the great telescope which can peer so far out into space and back into time will reveal the same density of nebulae with the same overall age-range and characteristics as, we can see near by. Bondi was quite clear about this.

It is the purpose of a scientific hypothesis to stick out its neck, that is, to be vulnerable. It is because the perfect cosmological principle is so extremely vulnerable that I regard it as a useful principle. It is something that could be 'shot down' by experiment and observation. . . . (Bondi, 'Rival Theories of Cosmology' from a broadcast talk. BBC, 1959.)

17

The Extraordinary Observations
of Recent Astronomy

This bold challenge was about to be taken up. There had been a comparative lull in new observations since the discoveries of Hubble and during this time cosmological speculation had gone almost unchecked. Now there was a new name in astronomy—Walter Baade—and the results he achieved were very fundamental although not always quite what the theorists had hoped for! He began by probing into the dusty centres of nearby galaxies using red-sensitive photographic film in a similar way to infra-red photography on earth, which can penetrate a foggy atmosphere. Here he found a different population of stars containing more old red-giants, clusters, and fast-moving stars which were rich in the heavier metal atoms gathered from the dusty background. The outer spiral arms of the galaxies, on the other hand, contained more patches of incandescent hydrogen gas out of which brilliant young blue stars seemed to be condensing. Cepheid variables occurred in both regions of the galaxy but those among the younger star population of the spiral arms were very much brighter than those closer to the centre. The Cepheid variables which we know from our own galaxy, like the Pole-Star, belong to this fainter category. But the Cepheids that Hubble had used to estimate the distances to outer galaxies had been the more prominent younger ones and so it followed that a systematic error had arisen in all his calculations. In 1952 Baade announced to the Astronomical Congress that the scale of the observable Universe must be at least twice as big as had been supposed (since these intrinsically brighter stars must be farther away). More recent estimates have made the distances bigger still. The red-shifts of the nebulae, and therefore their speeds, still held good but the age of the Universe became immediately far greater. The epoch when this continual expansion first began is now thought to be as long ago as 13,000 million years.

The Universe was now bigger, emptier, and older than ever before. In some ways this suited those cosmologies which had postulated a moment of creation for it resolved the conflict between the age of the Universe and that of the Earth. There was now plenty of time for the planets to have formed and for their radioactive atoms to have decayed. Also a new measure of time—the long, slow evolution of the galaxies themselves—now came under scrutiny.

THE LIFE AND DEATH OF STARS

The serene beauty of the great spiral galaxies, like the Andromeda nebula, is probably deceptive. The dust in their centres is evidence of millions upon millions of years of stellar violence including the dazzling disintegration of 'supernovae' whose climactic end we see so clearly even at great distances. This dust is composed of the heavier atoms which Steady State Theory had already suggested might be synthesised by the colossal heat of an exploding super-star. If we search for old stars which were formed early in the history of a galaxy we shall recognise them by the high velocities that they would have acquired over millions of years in this turbulent environment. Some of these stars have been thrown clear of the nuclear region and form a 'halo' region outside the disc of the galaxy. Their spectra show a great poverty of heavy atoms; only hydrogen and the helium formed from it, seem to be present—as if, in the days of their birth, the galaxy had not yet acquired its load of debris from supernovae eruptions. Since their path now takes them clear of the rich dust of the nucleus the composition of these halo stars is probably almost unchanged from the remote date of their condensation and is

197

evidence for the primeval nature of hydrogen, the lightest and simplest element known. This evidence seemed to tell strongly against Lemaître's theory of the complex original super-atom which has only subsequently been breaking down into the lighter elements. The protagonists of the Steady-State Theory could congratulate themselves on correctly predicting the original state of matter but

The 'Orion Nebulae'—a region of hot hydrogen gas in our own Galaxy, where new blue stars are being born

Baade's work had produced some indigestible items for them too.

Galaxies exist in many different shapes—elliptical, spherical, spiral, and irregular—and since the days of Hubble's first classification there had been speculation that this indicated a sequence in the ages of the galaxies. Now that observers know what to look for, a rather

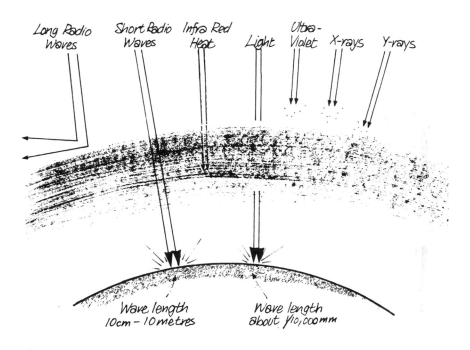

surprising result has emerged. The oldest stars, the metal-rich red-dwarfs with considerable movement, are to be found in *every* galaxy. The young, blue hot stars are not always present as some of the elliptical galaxies seem to have lost all their hydrogen 'source gas' by collision, or some other unknown process. But not one single example of a really young galaxy has been found. This might be evidence that all the galaxies began to condense from the primeval hydrogen at the same epoch, about ten thousand million years ago; it certainly casts grave doubt on the Steady State Theory of continuous creation. In fact there has been such a wealth of new observations during the last twenty years that cosmological speculation has been forced to take a back seat, at least temporarily. This is the time for absorbing and digesting the new information—it will be the task of the next generation of cosmological theory to explain these facts and fit them into a new scheme of the Universe.

ASTRONOMY USING OTHER RADIATIONS

Visible light is not the only carrier of messages that can cross the vast empty reaches of our curved and expanding Universe. There is a continuous spectrum of electromagnetic radiations ranging from radio waves more than a million times longer than light to X-rays and gamma-rays which are more than a million times shorter. Every shining radiant object in the heavens could well be transmitting energy at many different wavelengths. It is only by accident of biological evolution that our eyes respond to one small range of these waves which we call 'light'. Moths, for example, can see ultra-violet light which is quite black to us. If such cosmic radiations do exist, where are they? We know that the majority of such waves can never arrive on Earth at all because our atmosphere is a most effective shield. The dangerous waves at the short end of the spectrum, which are responsible for many of the hazards of nuclear radiation, are mercifully absorbed by the upper layers of our atmosphere. Here they spend their energy tearing the outer electrons away from the atoms of the air and creating a charged shell known as the 'ionosphere' and this, in turn, reflects back into space most of the longer radio waves which might otherwise be impinging on the Earth. Carbon dioxide and water vapour in the air are responsible for absorbing most of the long infra-red radiation. All of these waves might be carrying vital information about the nature of the stars and galaxies, but they can never penetrate the air and reach down

200

Artist's view of the proposed 120in Manned Orbiting Telescopes in space

to us on the surface of the Earth. Gradually during the last ten years a start has been made on a new kind of Space Astronomy for which radiation detectors have been launched in satellites and space-probes which rise high above our blanket atmosphere. Here, from their lofty vantage point it is hoped that these telescopes will build up for us a new picture of the radiating Universe painted in 'colours' that we cannot see of objects about which we may now know nothing.

RADIO-ASTRONOMY

However, there is one useful range of short radio waves which do pierce through the atmosphere and can be detected on Earth. It is from these radiations that most of the recent astonishing discoveries about the Universe have been made. Any hot object can emit radio waves, but in such small quantities that their study might well be supposed to bring little reward. Fortunately it turned out that there were far more effective transmitters than these in the skies.

Even when Radio-Astronomy began it was not our blazing sun which was the first source of 'radio noise' to be detected. In 1935 Karl Jansky picked out the first cosmic 'hiss' through simple headphones from a rotating aerial in his back garden, and he found that it reached a peak when the aerial was directed towards the distant centre of our galaxy and not when it was pointing directly at the Sun. During the war great advances were made in wireless and radar technology and when it was over, huge radio telescopes were designed like the 'Great Ear' at Jodrell Bank and the arrays of field aerials at Cambridge. Such receivers are so sensitive that they can pick up even the faint continuous whisper of radio waves that our own warm earth gives out; but still the 'brightest' radio objects in the sky were rarely the simple hot bodies that the optical telescope picks out. Indeed the success of Radio-Astronomy is that it has made 'visible' to us thousands of strange celestial bodies which might otherwise have gone unnoticed among their more optically spectacular neighbours.

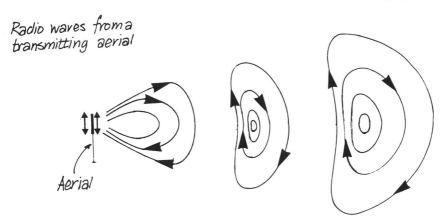

Radio waves from a transmitting aerial

Aerial

The closed loops of magnetic field form part of the wave that is radiated at the speed of light

Our own wireless transmitters generate waves by making electrons oscillate up and down an aerial. In this way they make their own fluctuating magnetic field with which they react to send out electromagnetic waves. These travel with the speed of light as Hertz discovered a century ago with his primitive spark oscillators. When-

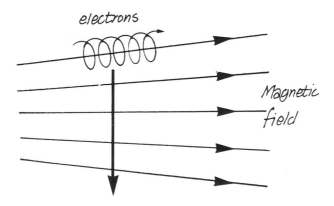

electrons

Magnetic field

Radio waves from an electron spiralling in a magnetic field

ever charged atoms or electrons spiral or vibrate in a strong magnetic field they too will generate energy in the form of radio waves which our radio-telescopes can pick up. The strongest single source of radio-waves in the sky is the Crab Nebula, the still expanding remnants of a supernova within the Milky Way which erupted in 1054. This hot exploding gas contains free electrons which spiral round the radiating lines of magnetic force and send out copious radio signals. The centre of our galaxy is also a turbulent region where free electrons are oscillating in powerful magnetic fields so that it, too, is bright with radio noise. Neighbouring galaxies have been shown to generate their own radio waves in the same way although their great distance from us dims the strength of their signals. But this is, perhaps, the least exciting feature of the new Astronomy.

The Crab Nebula has been pin-pointed as the remains of a 'super-Nova' which exploded in AD 1054. It emits both radio waves and X-rays and the strength of its magnetic field and spiralling electrons is such that even the visible light it gives out is polarised

Neutral hydrogen gas can also generate a radio wave when its cold molecules collide in space. This is energy at a quite definite wavelength—21cm—as specific to hydrogen as yellow coloured light is to sodium. By 'tuning in' to this frequency radio-telescopes have confirmed the abundance of this life-giving gas within the spiral arms of our galaxy where Baade's 'new population' of blue hot stars are being born. In the older central region of our galaxy there is, by contrast, very little hydrogen although it is radiating loudly at a whole range of other frequencies that no interstellar dust can block out from us. Indeed it is the elliptical galaxies which have no spiral arms and little free hydrogen like our own nucleus that are the strongest radiators of radio waves. If Baade was right it seems that within galaxies it is the birth and youth of stars which are comparatively peaceful and old age which brings about turbulence and strife!

FINDING THE 'RADIO GALAXIES'

Radio maps of the heavens showed many 'bright' objects which remained unidentified for several years. Only when radio astronomers had acquired the skill to pin-point these objects with real accuracy could the great optical telescopes be brought into the hunt. In 1951 Walter Baade trained the 200-inch reflecting telescope on Mount Palomar towards the exact patch of sky from which the radio-astronomers had been receiving very strong signals and eventually, using the most sensitive photographic film and an exposure of several hours, he discovered a faint but curious object in the middle of his plate. This is one of the most dramatic and exciting photographs of modern astronomy. At the precise location given by the radio signals lay a remote 'dumb-bell object'—a double galaxy more than 700 million light-years away. It seemed extraordinary that the second brightest radio object in the sky should be at such a remote distance from us, but subsequent discoveries showed that other 'radio-galaxies' were even farther away, some as far as 5,000 million light-years! At first it was supposed that such vast quantities of radio energy must be due to a collision between two galaxies but when some other photographs showed erupting filaments of gas from such radio sources it began to seem more likely that the burst of radio waves was due to a kind of violent galactic explosion. There is a

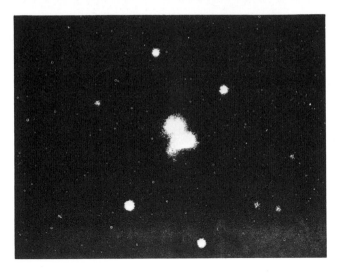

Above: Baade's 'Dumb-bell'. The first radio object to be seen. Cygnus 'A' radio source. Below: A remote radio galaxy. Collision or explosion? M 87 jet

mystery here about the lives of galaxies hundreds of millions of years ago that cosmology can still only wonder at.

These radio galaxies are quite plentiful in the Universe. Only a few of them can be picked out by optical telescopes near the extreme limit of their power. The others must be presumed to lie still farther away giving the new radio-telescopes a longer view of the Universe, in some respects, than the best of the conventional instruments. Once this was realised counts of these radio sources were undertaken at Cambridge in order to make a test between the two main rival theories of cosmology. In 1961 Martin Ryle announced the results of his first careful sampling of radio sources at different distances: *There were more faint radio sources far out in space than there were near to us.* If this is true it must mean that long ago, when these waves were first emitted, radio galaxies were more common than they are now. Perhaps they represent an early stage in the growth of galaxies. At all events this result was a heavy blow to the Steady

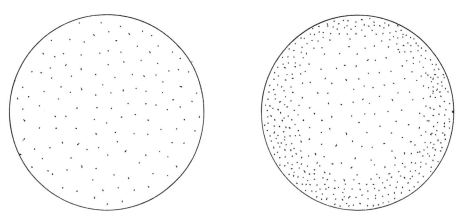

Uniform distribution of the galaxies on the Steady State Theory. Greater density of radio galaxies at greater distances as actually counted. (Reminiscent of Milne's prediction from his time theory)

Cambridge University radio telescope with Professor Sir Martin Ryle, its director and now the Astronomer Royal

State Theory for it seemed to prove that the Universe has not always been as it is now and that some kind of evolution in time may have taken place.

THE MYSTERY OF 'QUASARS'

During the last ten years even more curious radio sources have been detected. These are 'point sources' as sharp and clear as the bright point image of a single star. For this reason they were named 'quasi-stellar objects' which was quickly shortened into 'quasars'. A few of these have now been identified optically and found to be very bright blue hot stars whose light output fluctuates noticeably over a period of a few seconds or days. When photographs of their spectra were obtained they were found to exhibit an astonishingly large 'red-shift' in the colours of their emission lines. This would, by Hubble's law, correspond to distances so immense that their light may have been travelling towards us since the earliest years of the Universe! Could these quasars really date back to that remote epoch immediately after the first Big Bang? There are several other puzzling features about these strange radio stars. It seems that they are all strong emitters of ultraviolet light which may indicate enormously high temperatures due to some new kind of internal energy. Stranger still, their light shows dark absorption lines (due perhaps to a cooler envelope of gas around them) which has a different, smaller red-shift as though it is receding from us at a slower speed than that of the swift quasar itself! The whole subject of quasars is still an unsolved riddle which is enormously intriguing.

'CURIOUSER AND CURIOUSER'

After the last half-century of advance and speculation how does this great Universe appear to us now? Amid all the rival and uncertain theories one idea alone stands out as a new key to understanding,

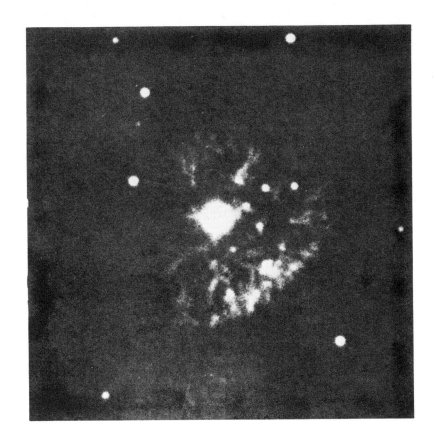

Example of a quasar with a shell-like structure. The central star and its surrounding cloud of hot gas may be the result of some violent stellar explosion

that the great twin concepts of space and time are linked indissolubly together, not only by the logic of Relativity but also by the constant speed of light as it travels towards us carrying knowledge of worlds which had their being both far away and long ago. Many centuries passed before man could accept the reality of a great void in which the stars and galaxies exist, but for us now this emptiness is pierced and irradiated by millions of waves carrying tantalising and only half-decipherable information about the remote bodies within this empty gulf. Perhaps the most ironic aspect of our situation is that now, when an attempt is made to interpret these observations, sophisticated man stands alongside his primitive forebears trying to construct a theory about the Beginning of Time which, despite its advanced mathematics, shows a curious similarity to all the ageless myths of Creation. In one way, however, the observation of science has transcended the simple gaze of a child. It seems now that the serene tranquillity of a star-lit night may be a misleading clue to the nature of the cosmos. Every new advance records happenings of almost inconceivable violence throughout the Universe. Consider the four following categories of modern mysteries:

(a) Cosmic Rays. These are high-energy atomic particles which bombard our atmosphere continuously at speeds only a fraction below that of disembodied energy like light. Where do they come from?

(b) Gravitational Waves are almost imperceptible because they are so elusive but now that we can detect them it has been shown that millions of stars would have to be totally annihilated to produce such energy. How can this be?

(c) Radio galaxies are unbelievably loud when their great distances from us are considered. If our photographs are to be believed then whole galaxies like these are actually exploding. What could possibly cause violence on such a vast scale?

(d) Quasars defy imagination! If they really are at the immense distances from us that their 'red-shifts' suggest they must be both brighter and more massive than a thousand galaxies and yet pulsating as rapidly as a single star. Can anyone believe so wild a story?

Those who fear that probing into space and time with the 'cool' instruments of science might destroy the romance and mystery of the Universe have little cause to worry. The science of cosmology is a challenge to more than just man's technology and his intellectual powers of comprehension—the simple power of his imagination is also stretched to the limit by the study of space and time as it has always been throughout the ages.

Select Reading List

So many books are published each year on scientific subjects that it is quite impossible to be familiar with them all. However, it is hoped that the following list of books may prove a useful jumping-off point for those whose interest has been aroused. None of them are mathematical nor technical in their treatment, many are very well illustrated and some are positively entertaining. I have enjoyed reading them all, but tastes differ and there are many others from which to choose.

Abell, G. *Exploration of the Universe* (1970)
Asimov, I. *Environments Out There* (1968)
BBC Programme Talks. *Einstein, the Man and His Achievement* (1967)
Berlage, H. P. *The Origin of the Solar System* (1968)
Bernardini and Fermi. *Galileo and the Scientific Revolution* (1963)
Bondi, Bonnor, Littleton and Whitrow. *Rival Theories of Cosmology*
Bondi, H. and others. *Pioneering in Outer Space*
Born, M. *Physics in my Generation* (1970)
Brumbaugh, R. S. *The Philosophers of Greece* (1966)
Caldar, N. *Radio Astronomy*
Crombie, A. C. *Augustine to Galileo* (Vol 2) (1969)
Einstein, A. *The World as I See It*
Farrington, B. *Science in Antiquity* (1969)
Gamow, G. *Gravity* (In the Science Study series) (1962)
Gamow, G. *Mr Tompkins in Paperback* (1965)
Harré (ed). *Early Seventeenth Century Scientists* (Gilbert, Bacon, Galileo, Kepler, Harvey, Van Helmont, Descartes) (1965)
Hoyle, F. *Galaxies, Nuclei and Quasars* (1966)
Hoyle, F. *The Nature of the Universe*

Kearney, H. *Science and Change 1500–1700*
Kendall, J. *Michael Faraday* (1955)
Koestler, A. *The Sleepwalkers* (1970)
Koestler, A. *The Watershed* (A biography of Kepler) (1961)
Kopal, Z. *Telescopes in Space* (1968)
Lovell, B. *Our Present Knowledge of the Universe* (1967)
More, Trenchard L. *Isaac Newton*
Newton, Isaac. *Opticks*
Ronchi, V. *The Nature of Light* (1970)
Sciama, D. *The Unity of the Universe* (1959)
Schatzman, E. L. *The Structure of the Universe* (1968)

Acknowledgements

Acknowledgement is due to the following for kindly allowing the illustrations to be used. The numbers refer to the page on which the illustration appears.

18 from *Science & Change*, World University Press; 32, 37, 39, 41 Science Museum, London; 67 from *History of the Royal Society*, 1667 by Thomas Sprat; 75 from *Opticks*, Dover Edition; 77 from *A Text Book on Light* by A. W. Barton, Longmans; 92 from *Light* by C. B. Daish, Eng Univ Press; 108 Royal Institution, London; 118 from *Scientific American*, Nov 1964, Prof R. S. Shankland; 134 from *Exploration of the Universe* by G. Abell, Holt Rinehart & Winston; 138 Science Museum, London; 165 from *Scientific American*, 1971; 172, 177, 180, 198 Science Museum, London; 201 from *Telescopes in Space* by Kopal, Faber, and the Boeing Company Aerospace Division; 204 from *Scientific American*, August 1966; 206 both from *Galaxies, Nuclei and Quasars* by Fred Hoyle, Heinemann; 208 Professor Sir Martin Ryle; 210 from *Scientific American*, December 1970.

Index